Fundamentals of Magnetism

Fundamentals of Magnetism

Mario Reis

Instituto de Física
Universidade Federal Fluminense
Niterói, Rio de Janeiro
Brazil

AMSTERDAM • BOSTON • HEIDELBERG • LONDON • NEW YORK
OXFORD • PARIS • SAN DIEGO • SAN FRANCISCO • SINGAPORE
SYDNEY • TOKYO
Academic Press is an imprint of Elsevier

Academic Press is an imprint of Elsevier
The Boulevard, Langford Lane, Kidlington, Oxford OX5 1GB, UK
Radarweg 29, PO Box 211, 1000 AE Amsterdam, The Netherlands
225 Wyman Street, Waltham, MA 02451, USA
525 B Street, Suite 1900, San Diego, CA 92101-4495, USA

First edition 2013

Notice
No responsibility is assumed by the publisher for any injury and/or damage to persons or property
as a matter of products liability, negligence or otherwise, or from any use or operation of any meth-
ods, products, instructions or ideas contained in the material herein. Because of rapid advances in
the medical sciences, in particular, independent verification of diagnoses and drug dosages should
be made.

Library of Congress Cataloging-in-Publication Data
Reis, Mario, 1976–
    Fundamentals of magnetism/by Mario Reis. — First edition.
        pages cm
    Includes bibliographical references and index.
    ISBN 978-0-12-405545-2 (alk. paper)
1. Magnetism—Textbooks.   I. Title.
QC753.2.R45   2013
538—dc23
                                                          2012044516

British Library Cataloguing in Publication Data
A catalogue record for this book is available from the British Library

ISBN: 978-0-12-405545-2

For information on all Academic Press publications
visit our website at store.elsevier.com

Working together
to grow libraries in
developing countries

ELSEVIER   Book Aid
            International

www.elsevier.com • www.bookaid.org

To my guides

Contents

Preface

This book is a consequence of (almost) twenty years working with magnetism and magnetic properties of materials. Of course, there are a plenty of books on magnetism; but a new book always help students, since further examples, text and figures can emphasize and guide the learning process.

The target audience is undergraduate and graduate students, mainly those who want to follow the carrier on magnetism research. This book can also be useful for researchers of other areas, like chemistry and engineering, that need a consult and fast look on the subject.

After study on this book, the student/researcher may be able to describe the basic difference between the magnetic orderings, namely dia-, para-, ferro-, antiferro-, and ferri-magnetic materials. In addition, the reader will also be able to identify and obtain important parameters from the experimental data (magnetization, susceptibility, and specific heat) of materials with those magnetic orderings.

This book was designed to be self-contained. In principle, students that already attended lectures on Quantum Mechanics (QM) and Statistical Mechanics (SM) will be able to understand the presented content; however, facing any difficulty, a brief survey on the fundamentals of QM and SM is given, focused on the themes needed to understand the magnetic properties described. This survey is also useful for students that never studied QM and SM, since it goes deeper enough to give the necessary background.

Thus, this book has three parts: the first one is a *Background* and discusses notation and important Hamiltonians contributions to the magnetism, as well as a review/ introduce QM and SM fundamentals. The second part deals with *Non-cooperative magnetism*, namely diamagnetism and paramagnetism, the most fundamental theories on magnetism. Finally, the third and last part deals with *Cooperative magnetism*, detailing the magnetic interactions, long-range interactions, and Landau theory (this last is provided to the reader to understand first- and second-order phase transitions). This third part ends with a modern subject: molecular magnetism and its applications. The book also contains exercises to promote practice to the reader.

Some important acknowledges: Alberto Passos Guimarães and Ivan Oliveira—authors of some books on Physics and examples to me; Vitor Bastos, Stephane Soriano, Sergio Resende, Vitor Amaral, and Alexandre Carvalho for the careful reading of the text; FAPERJ, CAPES, CNPq, and PROPPi-UFF for the financial support; and for the Elsevier editorial team for the efficient support. Finally, a special acknowledgment to my family, always present in my life.

Finally, few words on me: I was born in 1976, at Rio de Janeiro, Brazil and concluded my bachelor's degree in Physics in 1997 at Federal University of Rio de Janeiro (UFRJ). My master's (2000) and Ph.D (2003) degrees in Physics, more precisely on Magnetism and Magnetic Materials (experimental works), were obtained at the Brazilian Center for Research on Physics (CBPF), with a traineeship at the University of Aveiro (Portugal). I then concluded two post-doctoral programs in 2006, also at University of Aveiro, and then got a position as researcher in the same university, until 2010, when I moved back to Brazil. Since then, I am professor of physics at Federal Fluminense University (UFF), at Rio de Janeiro. At the present date, I have written more than 70 scientific papers on Magnetism and an other book, entitled Magnetismo Molecular (in Portuguese). I hope my experience on this subject, as well as my close contact with students, will help to provide to the community a useful text on this interesting subject.

Questions, comments, or criticisms are welcome. Do not hesitate to contact me: marior@if.uff.br

Mario Reis
Rio de Janeiro, Brazil
January 2013

List of Figures

List of Tables

Part One
Background

1 Introduction

1.1 Quantities and units

There are only a few quantities to be defined on magnetism; and some equations depend on the unit system chosen (thus, attention on this issue is needed). The most common systems are: SI (International System) and CGS (Centimeter-Gram-Second).

Let us start these definitions from the magnetic field that exists when a force $\vec{F}$ acts on a test particle with electrical charge q and velocity $\vec{v}$. This force, known as Lorentz force, is:

$$\vec{F} = q\left(\vec{E} + \vec{v} \times \vec{B}\right) \quad \text{[SI]}, \tag{1.1}$$

$$\vec{F} = q\left(\vec{E} + \frac{\vec{v}}{c} \times \vec{B}\right) \quad \text{[CGS]}, \tag{1.2}$$

where $\vec{B}$ is the magnetic induction created by the source of magnetic field $\vec{H}$ and $\vec{E}$ the electrical field (for the sake of simplicity, this last will not be considered). In the vacuum, these two quantities are related:

$$\vec{B} = \mu_0 \vec{H} \quad \text{[SI]} \tag{1.3}$$

and the constant of proportionality is the vacuum magnetic permeability μ_0.

Considering the International System (SI), the magnetic induction B is measured in Tesla (T) and corresponds to the force of 1 Newton (N) under an electrical charge of 1 Coulomb (C), that moves with 1 meter/second (m/s), in a direction perpendicular to B ($\vec{v} \perp \vec{B}$). The magnetic field H is measured in Ampere/meter (A/m), and the vacuum magnetic permeability μ_0 assumes the value of $4\pi \times 10^{-7}$ Henry/meter (H/m). On the other hand, considering the CGS system, the quantities have the same meaning, however, different values and units: $\mu_0 = 1$ (dimensionless), the magnetic induction B is measured in Gauss (G) and the magnetic field H measured in Oersted (Oe). Thus, the above equation can be rewritten as

$$\vec{B} = \vec{H} \quad \text{[CGS]}. \tag{1.4}$$

As an example, let us obtain the magnetic induction $\vec{B}$ and the magnetic field $\vec{H}$ due to a conducting ring of radius a and with an electrical current I, as depicted in

Fundamentals of Magnetism. http://dx.doi.org/10.1016/B978-0-12-405545-2.00001-1

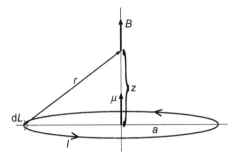

Figure 1.1 Magnetic induction $\vec{B}$ due to a ring with electrical current.

Figure 1.1. To this purpose, we must use the Biot-Savart law:

$$d\vec{B} = \frac{\mu_0 I}{4\pi} \frac{d\vec{L} \times \vec{r}}{r^3} \quad [\text{SI}] \tag{1.5}$$

that gives us the elemental magnetic induction $d\vec{B}$ corresponding to the elemental ring segment $d\vec{L}$ with current I. In addition, $\vec{r}$ represents the distance between $d\vec{L}$ and the point to obtain $d\vec{B}$. Thus, for the geometry proposed in Figure 1.1, the magnetic induction at a point z above the ring main axis is:

$$\vec{B} = \frac{\mu_0 I}{2} \frac{a^2}{(z^2 + a^2)^{3/2}} \hat{k} \quad \text{and} \quad \vec{H} = \frac{I}{2} \frac{a^2}{(z^2 + a^2)^{3/2}} \hat{k}. \tag{1.6}$$

Still considering the example of Figure 1.1, it is possible to define the magnetic moment of the system:

$$\vec{\mu} = I \mathcal{A} \hat{n}, \tag{1.7}$$

where $\mathcal{A} = \pi a^2$ corresponds to the ring area and $\hat{n}$ an unitary vector normal to the ring. The magnetic moment μ is then measured in Am^2 (SI), or erg/G (CGS). It is important to define that erg/G = emu (*electromagnetic units*, quite common in the literature, mainly to express experimental results).

Let us now consider other way to produce a magnetic field, i.e., using a permanent magnet with a magnetic moment $\vec{\mu}$. For values of $\vec{r}$ far from the magnet, the magnetic field produced is the same of that produced by the ring with the same magnetic moment $\vec{\mu}$ (previous discussion). Figure 1.2 compares these two situations. The magnetic field due to the magnet is then given by the dipolar expression:

$$\vec{H}(\vec{r}) = \frac{1}{4\pi} \left[\frac{3(\vec{\mu} \cdot \vec{r})\vec{r}}{r^5} - \frac{\vec{\mu}}{r^3} \right] \quad [\text{SI}]. \tag{1.8}$$

We leave as an exercise to the reader to show that Eqs. (1.6) and (1.8) are the same for a point far from the magnetic moment.

Let us now consider an ensemble of magnetic moments inside a volume V. The volume magnetization can then be defined as:

$$M = \frac{\sum_i \mu_i}{V}. \tag{1.9}$$

The magnetization M is measured as A/m (SI), or emu/cm^3 (CGS). Analogously, it is possible to define molar and mass magnetization.

For material media with magnetization $\vec{M}$, the magnetic induction inside the material assumes the form:

$$\vec{B} = \mu_0(\vec{H} + \vec{M}) \quad \text{[SI]}. \tag{1.10}$$

$$\vec{B} = \vec{H} + 4\pi\vec{M} \quad \text{[CGS]}. \tag{1.11}$$

A magnetized material, i.e., when the magnetic moments are (either partially or totally) aligned in one specific direction, creates "free poles" on the surface of the material, as shown in Figure 1.3. These "poles" create an additional field $\vec{H}_d$ in opposition

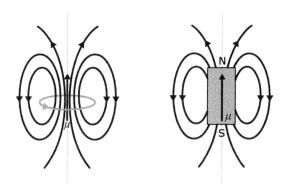

Figure 1.2 Magnetic field due to a (left) ring and (right) permanent magnet.

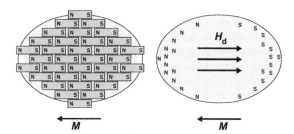

Figure 1.3 The orientation of the magnetic moments of the material creates "free poles" on the surface. This effect is the origin of the demagnetization field inside magnetic materials; that, on its turn, is contrary to the magnetization.

(a) **(b)**

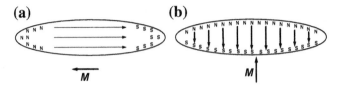

Figure 1.4 Scheme of the demagnetization field for an ellipsoid magnetized along the (a) longer and (b) smaller axis. Intuitively, the value of H_d is bigger for the case (b), since there are much more "free" poles and those are closer.

to the magnetization $\vec{M}$ and then, inside the magnetic material, the magnetic field $\vec{H}$ must be rewritten as:

$$\vec{H} = \vec{H}_0 - \vec{H}_d, \tag{1.12}$$

where $\vec{H}_0$ represents the external applied magnetic field. However, $\vec{H}_d$, known as demagnetization field, is related to the magnetization in a simple fashion:

$$\vec{H}_d = N_d \vec{M}, \tag{1.13}$$

where N_d represents the demagnetization factor, related to the geometry of the material and the direction of $\vec{M}$. These relations are of easy understanding. Image a magnetic material with ellipsoidal shape and magnetization along the long axis, as shown in Figure 1.4a. Those two groups of "free poles" are distant of each other and then the demagnetization factor N_d is smaller compared to the case of Figure 1.4b, where the magnetization is along the short axis. For this last case, those two groups of "free poles" are closer and, consequently, the demagnetization factor N_d is bigger. In what concerns dimensions, to obtain the value of the demagnetization factor in the CGS system, we must divide this value in the International System by 4π, i.e.,

$$N_d[\text{CGS}] = N_d[\text{SI}]/4\pi. \tag{1.14}$$

It is important to stress that, in the International System, $0 < N_d < 1$. Table 1.1 summarizes some demagnetization factors, for some different geometries and directions of the magnetization.

To be rigorous, the demagnetizing factor, as well as other quantities in magnetism, is a tensor.[1] For our case, the demagnetization factor is a *rank-2* tensor, i.e., $n = 2$, with dimension $m = 3$ (due to the three axes: x, y, and z). Thus, we can write the demagnetizing field as:

$$\begin{pmatrix} H_{d,x} \\ H_{d,y} \\ H_{d,z} \end{pmatrix} = \begin{pmatrix} N_{xx} & N_{xy} & N_{xz} \\ N_{yx} & N_{yy} & N_{yz} \\ N_{zx} & N_{zy} & N_{zz} \end{pmatrix} \begin{pmatrix} M_x \\ M_y \\ M_z \end{pmatrix}. \tag{1.15}$$

---

[1] A mathematical object with dimension m and n indexes; thus, m^n elements. Tensors are generalizations of a scalar (there is no index), and vectors (have one index).

Table 1.1 Demagnetization factor N_d for different geometries and direction of the magnetization. To obtain the corresponding value in the CGS, divide the values of the SI by 4π.

Geometry	Direction of the Magnetization M	Demagnetization Factor [SI]
Sphere	Isotropic	1/3
Plan	$\perp$ Plan	1
Plan	$\parallel$ Plan	0
Cylinder $(l/d = 1)^a$	$\parallel l$	0.27
Cylinder $(l/d = 5)$	$\parallel l$	0.04
Needle (cylinder with $l \gg d$)	$\parallel l$	0
Needle	$\perp l$	1/2
Toroid	$\parallel$ Axis	0

^{a} l: length, d: diameter.

To better understand the scenario above described, let us consider a magnetized sphere, with magnetization $\vec{M}$ induced by a magnetic field $\vec{H}$. In the SI, out of the sphere, the magnetic induction can be given by Eq. (1.3); while inside the material we must consider Eq. (1.10). However, inside the magnetic material, the magnetic field will be reduced due to the demagnetizing field; and then, if we carry Eqs. (1.13) and (1.12) into Eq. (1.10), we find, for the magnetic induction inside the material:

$$\vec{B} = \mu_0[\vec{H}_0 + \vec{M}(1 - N_d)] \quad [\text{SI}]. \tag{1.16}$$

Analogously, for the CGS system:

$$\vec{B} = \vec{H}_0 + \vec{M}(4\pi - N_d) \quad [\text{CGS}]. \tag{1.17}$$

Figure 1.5 clarifies this scenario.

However, how can the demagnetizing field be obtained? For the case in which the magnetization is uniform and there are no external currents on the magnetic material, the magnetic field inside this material can be evaluated from:

$$\vec{H}_d = \frac{1}{4\pi} \oint_S (\vec{M} \cdot \hat{n}) \frac{\vec{r}}{r^3} dS \quad [\text{SI}], \tag{1.18}$$

where dS represents a surface element, $\vec{r}$ the position vector from the origin of the coordination system to the integration point, and $\hat{n}$ an unitary vector normal to the surface of the material. We leave as an exercise to obtain, from the above equation, the demagnetizing field (and factor), due to a magnetized sphere.

Finally, let us define the magnetic susceptibility (definition that does not depend on the unit system used, i.e., it works for both, SI and CGS):

$$\chi = \lim_{H \to 0} \frac{\partial M}{\partial H}. \tag{1.19}$$

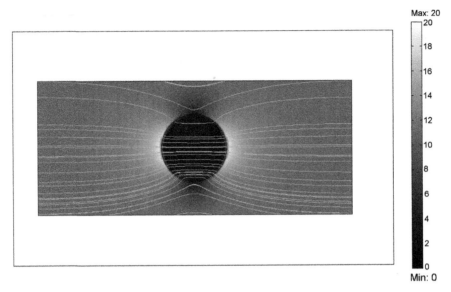

Max: 20

Figure 1.5 This figure represents an uniform magnetic field applied to a magnetic material. Those continuum lines are the field lines, deformed due to the presence of the magnetic material (central sphere). The gray scale on the right (where the number are in arbitrary units) represents the field intensity. Note that inside the magnetic material the field is less intense, due to demagnetization field. This figure was done after a simple simulation (not described in this book), with the FEMLAB program.

For small values of the magnetic field H, we can consider, at a first approximation, $M = \chi H$, and then Eqs. (1.10) and (1.11) can be rewritten as

$$\vec{B} = \mu_m \vec{H}, \tag{1.20}$$

where

$$\mu_m = \mu_0(1 + \chi) \quad \text{[SI]}, \tag{1.21}$$
$$\mu_m = 1 + 4\pi\chi \quad \text{[CGS]} \tag{1.22}$$

represent the magnetic permeability of the material.

To conclude this section, Table 1.2 summarizes the main quantities defined for both, SI and CGS systems of units.

1.2   Types of magnetic arrangement

In this section, let us discuss, qualitatively, some magnetic arrangements. By means of intuitive arguments, the idea is to predict the behavior of the magnetization as a function of both, magnetic field and temperature, as well as the inverse magnetic susceptibility as a function of temperature. We can start discussing the concept of (non)cooperativity.

Table 1.2 Main quantities on magnetism. The relation between both systems (CGS and SI), is: [CGS] = (conversion factor) × [SI].

Quantities	Symbol	CGS	Conversion Factor	SI	Obs.
Magnetic induction	B	G	10^{-4}	T	a
Magnetic field	H	Oe	$10^3/4\pi$	A/m	b
Magnetic moment	μ	erg/G ($\equiv$ emu)	10^{-3}	Am^2	
Volume magnetization	M	emu/cm^3	10^3	A/m	
Mass magnetization	M	emu/g	1	Am^2/kg	
Volume susceptibility	χ	dimensionless	4π	dimensionless	
Mass susceptibility	χ	emu/gOe	$4\pi \times 10^{-3}$	m^3/kg	
Magnetic permeability	μ_m	dimensionless	$4\pi \times 10^{-7}$	H/m	c
Demagnetization factor	N_d	dimensionless	$1/4\pi$	dimensionless	

[a] G: Gauss, T: Tesla.

[b] Oe: Oersted, A/m: Ampere per meter.

[c] H/m: Henry per meter. In the vacuum, $\mu_m = \mu_0 = 1$ [CGS] and, consequently, $\mu_m = \mu_0 = 4\pi \times 10^{-7}$ H/m [SI].

Magnetization is noncooperative when it is ruled only by the magnetic behavior of the individual magnetic moments, i.e., there is no interaction between those. On the other hand, cooperative ordering depends on the interactions between the numerous (1 mol), magnetic moments and, if we imaginarily remove one magnetic moment from the bulk, this unique one is no longer able to describe the magnetic behavior of the ensemble. There are basically two noncooperative orderings: diamagnetism and paramagnetism—the most fundamental concept of magnetism. However, for cooperative orderings, there is a lot of different arrangements: ferromagnetism, antiferromagnetism, ferrimagnetism, and many others. From now on, let us qualitatively describe those, one by one.

Diamagnetism: It is intrinsic to all of the materials. Generally speaking, when electrons are under an external applied magnetic field, the precession around the nucleus changes the frequency to promote an extra magnetic field and shield the external one. We then expect a linear and negative magnetization as a function of the magnetic field. This behavior does not depend (in a first approximation), on the temperature and then both, magnetization and susceptibility as a function of temperature assume a constant behavior. Figure 1.6 clarifies these words.

Paramagnetism: This ordering came from the polarization of the magnetic moments due to the action of a magnetic field. Let us consider an ensemble of magnetic moments at a certain temperature. Without applied magnetic field, there are magnetic moments pointing toward all of the directions and then the magnetization is zero. Application of a magnetic field promotes a relative orientation of the magnetic moments, increasing the value of the magnetization. At (quite) high values of magnetic field, all of the magnetic moments are parallel to each other, and the magnetization reaches its maximum value (saturation: M_s).

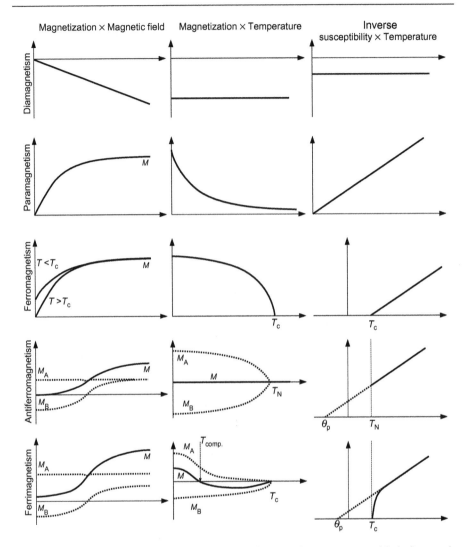

Figure 1.6 Qualitative scheme of the magnetic orderings. Each row means one kind of magnetic ordering and each column represents one thermodynamic quantity. The origin of the axes is the abscissa-ordinate crossing point.

For a certain value of applied magnetic field, decreasing temperature, this relative orientation of the magnetic moments increases, increasing therefore the magnetization and, at 0 K, the magnetization reaches its maximum value. The magnetic susceptibility, obtained for low values of applied magnetic field, is given by the Curie law:

$$\chi = \frac{C}{T},$$ (1.23)

where C is the Curie constant. The inverse magnetic susceptibility is then a straight line; and the zero $1/\chi$ is found at 0 K (it means that there are no interactions between the magnetic moments). Figure 1.6 clarifies these words.

Ferromagnetism: This kind of magnetic ordering belongs to the cooperatively ordered systems. One magnetic moment depends on the neighbors to then create the magnetic ordering: it is a long-range interaction. As mentioned before, if we imaginarily remove one magnetic moment from the system, this one will no longer be magnetic as the bulk, since this needs the neighbors. It is like a house of cards; one card needs the neighbor and then the house only exists due to a cooperative support. If we remove one card (supposing the house keeps on), this unique one is no longer a house (but the house is still a house).

Two parameters characterize the ferromagnetic ordering: (1) the critical temperature T_c (named Curie temperature), above which the system behaves like a paramagnetic system, i.e., zero magnetization for zero applied magnetic field. Below T_c the system has spontaneous magnetizations, i.e., finite magnetization even for zero applied magnetic field. In a first approximation, T_c is a measure of how strong is the interaction between magnetic moments. (2) The saturation value of the magnetization M_s, analogously to the paramagnetic case, i.e., it measures the arithmetic sum of all of the parallel magnetic moments.

We then expect for the magnetization as a function of temperature curve a finite value of magnetization that decreases by increasing the temperature, up to a critical value T_c, above which there is no longer spontaneous magnetization. Concerning the magnetization as a function of magnetic field, there are two situations: the first one is for temperatures above T_c. For this case, as mentioned, the system behaves like a paramagnetic specie, and then, we expect a curve similar to the paramagnetic case. For temperatures below T_c, it has a spontaneous magnetization; that means the magnetization does not start from the origin of the system, but from a finite value of magnetization.

The interaction between magnetic moments creates an internal magnetic field, that acts on the own magnetic moments. This fact allows the system to reach the saturation value of the magnetization for relatively small values of magnetic field, in comparison to the paramagnetic case. See Figure 1.6 for further details. Finally, the magnetic susceptibility is given by the Curie–Weiss law:

$$\chi = \frac{C}{T - \theta_p},$$
(1.24)

where θ_p is the paramagnetic Curie temperature (for some models, $\theta_p = T_c$). The inverse magnetic susceptibility is also a straight line, and zero $1/\chi$ meets $T = \theta_p > 0$.

Antiferromagnetism: This kind of cooperative ordering is understood considering two magnetic sublattices: M_A and M_B, of the same magnitude. Each one is ferromagnetic and behaves (approximately) like the description mentioned above. The difference is that M_A is oriented in opposition to M_B; in other words: $M_A = -M_B$. The critical temperature below which these sublattices are spontaneously ordered is named Neel temperature T_N. Note that this ordering is not a simple addition of two ferromagnetic sublattices, aligned in an antiparallel fashion; there is an interaction between these two sublattices, making this system a bit more complex than these few words.

In what concerns the behavior of the magnetization as a function of magnetic field, at $T \ll T_N$, one sublattice (say, M_A) is aligned with the external magnetic field and then does not change by increasing field. The other sublattice (M_B) is in opposition to the field and then will be flipped due to the increasing of the field. The total magnetization is a sum of the sublattices ($M = M_A + M_B$). Figure 1.6 clarifies these words. The behavior of the magnetization as a function of temperature is built in a similar fashion as before: two sublattices working in opposition. For the case without external applied magnetic field, each sublattice has a ferromagnetic-like dependence with temperature and the total magnetization is then zero. Finally, the magnetic susceptibility is also given by the Curie-Weiss law, however, zero $1/\chi$ meets $T = \theta_p < 0$.

Ferrimagnetism: This cooperative ordering is quite similar to the previous one, however, for the present case, these two sublattices have different values of magnetic moment (i.e., $M_A \neq M_B$), still in opposition. The behavior of the magnetization as a function of magnetic field (for low values of temperature) is the same as before, however, due to the difference in the values of the magnetic moment of each sublattice, it is not zero at zero magnetic field. See Figure 1.6 for a better understanding. In what concerns the magnetization as a function of the temperature, since these two sublattices are different, these can cross themselves for a certain value of temperature T_{comp}, and then promote a compensation, where the total magnetization is zero. The system loses the spontaneous ordering above T_c, analogously to the ferromagnetic case. Finally, the magnetic susceptibility only follows the Curie-Weiss law for very high temperature. Close to T_c, the inverse magnetic susceptibility loses its linearity and assumes a hyperbolic-like behavior, with a downturn to zero $1/\chi$.

2 Hamiltonian of an Electron Under an Electromagnetic Field

The aim of this chapter is to provide the Hamiltonian terms important to develop the theoretical foundations of magnetism, namely the Zeeman, spin-orbit, and diamagnetic contributions.

2.1 Classical approach

First, let us remember some important fundamentals of classical mechanics: Lagrangian and Hamiltonian formalisms, that are the roots of quantum mechanics. In few words, let us try to summarize the ideas of these formalisms; only the minimum required equations to understand the scope of the present section, i.e., to obtain the (classical) Hamiltonian of an electron moving under the influence of an electromagnetic field.

Thus, the Lagrangian is defined as the difference between the kinetic T and potential V energies:

$$\mathcal{L}(q_i, \dot{q}_i, t) = T - V \tag{2.1}$$

and, in addition, it is written in terms of generalized coordinates q_i and generalized velocities $\dot{q}_i$, where the dot above q_i means the first and total derivative with respect to time t.

If we know the Lagrangian, it is possible to obtain the equation of motion for each generalized coordinate q_i of the system, through the Lagrange Equation:

$$\frac{\mathrm{d}}{\mathrm{d}t}\left(\frac{\partial \mathcal{L}}{\partial \dot{q}_i}\right) - \frac{\partial \mathcal{L}}{\partial q_i} = 0, \tag{2.2}$$

where

$$\frac{\mathrm{d}}{\mathrm{d}t} = \frac{\partial}{\partial t} + \sum_{i=1}^{N} \dot{q}_i \frac{\partial}{\partial q_i} + \sum_{i=1}^{N} \ddot{q}_i \frac{\partial}{\partial \dot{q}_i} \tag{2.3}$$

means the total time derivative. It is also possible to know the generalized (or conjugate) momentum of each generalized coordinate, and it is given by the equation:

$$p_i = \frac{\partial \mathcal{L}}{\partial \dot{q}_i}. \tag{2.4}$$

Fundamentals of Magnetism. http://dx.doi.org/10.1016/B978-0-12-405545-2.00002-3

From the Lagrangian formalism, it is possible to derive the Hamiltonian, which is defined as

$$\mathcal{H}(q_i, p_i, t) = \sum_{i=1}^{N} p_i \dot{q}_i - \mathcal{L} = T + V. \tag{2.5}$$

Lagrangian formalism works with $\{q_i, \dot{q}_i\}$ variables, while Hamiltonian formalism changes variables and works with $\{q_i, p_i\}$. The equation of motion of the system under consideration can also be obtained from the Hamiltonian Equations:

$$\frac{dq_i}{dt} = \frac{\partial \mathcal{H}}{\partial p_i} \quad \text{and} \quad \frac{dp_i}{dt} = -\frac{\partial \mathcal{H}}{\partial q_i}. \tag{2.6}$$

Thus, after this (really) brief survey on the Lagrangian and Hamiltonian formalisms, we may back to our main problem and derive the classical Hamiltonian of an electron traveling under an electromagnetic field. Let us therefore start from the Lorentz force:

$$\vec{F} = -e \left(\vec{E} + \vec{v} \times \vec{B} \right), \tag{2.7}$$

and write the electric and magnetic fields in terms of its potentials (Eqs. (2.A.7) and (2.A.11)):

$$m\ddot{\vec{r}} = -e \left(-\vec{\nabla}\phi - \frac{\partial \vec{A}}{\partial t} + \vec{v} \times (\vec{\nabla} \times \vec{A}) \right). \tag{2.8}$$

The second and third terms of the above parentheses can be written in a more convenient form. Second term: from Eq. (2.3) and considering $\vec{A} = \vec{A}(\vec{r}, t)$, the total time derivative can be written as:

$$\frac{d\vec{A}}{dt} = \frac{\partial \vec{A}}{\partial t} + (\vec{v} \cdot \vec{\nabla})\vec{A}. \tag{2.9}$$

Third term: from simple vectorial analysis, it is possible to write:

$$\vec{v} \times (\vec{\nabla} \times \vec{A}) = \vec{\nabla}(\vec{v} \cdot \vec{A}) - (\vec{v} \cdot \vec{\nabla})\vec{A}, \tag{2.10}$$

where

$$[(\vec{v} \cdot \vec{\nabla})\vec{A}]_i = \sum_{j} v_j \frac{\partial A_i}{\partial x_j} \quad (i = x, y, z). \tag{2.11}$$

Thus, replacing Eqs. (2.9) and (2.10) into Eq. (2.8) we obtain:

$$\frac{d}{dt}(-e\vec{A} + m\ddot{\vec{r}}) - \vec{\nabla}(-e\vec{v} \cdot \vec{A} + e\phi) = 0, \tag{2.12}$$

which can easily be written in terms of generalized coordinates q_i and generalized velocities $\dot{q}_i$:

$$\frac{d}{dt}(-eA_i + m\dot{q}_i) - \frac{\partial}{\partial q_i}(-e\dot{q}_i A_i + e\phi) = 0. \tag{2.13}$$

Now, it is time to compare the above equation with Lagrange Equation (Eq. (2.2)). Comparison of the first term of both equations yields:

$$\frac{\partial \mathcal{L}}{\partial \dot{q}_i} = -e A_i + m \dot{q}_i \tag{2.14}$$

and therefore

$$\mathcal{L} = -e \dot{q}_i A_i + \frac{1}{2} m \dot{q}_i^2 + \mathcal{C}_1, \tag{2.15}$$

where $\mathcal{C}_1$ *does not* depend on $\dot{q}_i$. On the other hand, comparison of the second term of Eq. (2.13) with its counterpart of Eq. (2.2) yields

$$\mathcal{L} = -e \dot{q}_i A_i + e \phi + \mathcal{C}_2, \tag{2.16}$$

where $\mathcal{C}_2$ *does not* depend on q_i. Now, from a simple inspection of Eqs. (2.15) and (2.16), we can find one of the possible solutions: $\mathcal{C}_1 = e \phi$ and $\mathcal{C}_2 = \frac{1}{2} m \dot{q}_i^2$. Thus, the Lagrangian of an electron moving under an electromagnetic field is:

$$\mathcal{L}(q_i, \dot{q}_i, t) = \frac{1}{2} m \dot{q}_i^2 - e \dot{q}_i A_i + e \phi. \tag{2.17}$$

Finally, let us now conclude our objective, that is to evaluate the Hamiltonian of an electron under an electromagnetic field. For this purpose, consider Eq. (2.5), where it is possible to obtain the Hamiltonian from the Lagrangian. Thus,

$$\mathcal{H} = p_i \dot{q}_i - \frac{1}{2} m \dot{q}_i^2 + e \dot{q}_i A_i - e \phi. \tag{2.18}$$

However, from Eqs. (2.4) and (2.14) it is possible to obtain the generalized momentum, that reads as

$$p_i = \frac{\partial \mathcal{L}}{\partial \dot{q}_i} = -e A_i + m \dot{q}_i \tag{2.19}$$

and then:

$$\dot{q}_i = \frac{1}{m} (p_i + e A_i) \tag{2.20}$$

and therefore substituting Eq. (2.20) into Eq. (2.18), the Hamiltonian resumes as (changing back to real coordinates, instead of those generalized):

$$\mathcal{H} = \frac{1}{2m} (\vec{p} + e \vec{A})^2 - e \phi. \tag{2.21}$$

2.1.1   Particular case: Uniform magnetic field

Let us now study the case where the magnetic field is uniform and points along the z direction, i.e., $\vec{B} = B\hat{k}$. We know that

$$\vec{B} = \vec{\nabla} \times \vec{A} = (\partial_y A_z - \partial_z A_y)\hat{i} + (\partial_z A_x - \partial_x A_z)\hat{j} + (\partial_x A_y - \partial_y A_x)\hat{k}, \quad (2.22)$$

where $\partial_{x_i} = \partial/\partial x_i$; and therefore, some different $\vec{A}$ vectors can produce $\vec{B} = B\hat{k}$. For instance, we can consider either

$$\vec{A} = A_y \hat{j} = Bx\hat{j} \qquad (2.23)$$

or

$$\vec{A} = A_x \hat{i} = -By\hat{i}. \qquad (2.24)$$

These gauges are known as Landau gauges. See Complement 2.A for further details. Another possibility is a linear combination of the above solutions:

$$\vec{A} = \frac{1}{2}B(-y\hat{i} + x\hat{j}). \qquad (2.25)$$

We also know that $\vec{r} \times \vec{B}$ can be written as (still considering $\vec{B} = B\hat{k}$):

$$\vec{r} \times \vec{B} = \begin{vmatrix} \hat{i} & \hat{j} & \hat{k} \\ x & y & z \\ 0 & 0 & B \end{vmatrix} = By\hat{i} - Bx\hat{j} \qquad (2.26)$$

and therefore we can rewrite the vector potential $\vec{A}$ in terms of the product $\vec{r} \times \vec{B}$:

$$\vec{A} = -\frac{1}{2}(\vec{r} \times \vec{B}). \qquad (2.27)$$

The Hamiltonian of Eq. (2.21) can then be rewritten as

$$\mathcal{H} = \frac{\vec{p}^2}{2m} - e\phi - \frac{e}{4m}\left[\vec{p} \cdot (\vec{r} \times \vec{B}) + (\vec{r} \times \vec{B}) \cdot \vec{p}\right] + \frac{e^2}{8m}(\vec{r} \times \vec{B})^2, \qquad (2.28)$$

where the above vector products can be summarized as[1]:

$$\vec{p} \cdot (\vec{r} \times \vec{B}) + (\vec{r} \times \vec{B}) \cdot \vec{p} = -2\vec{L} \cdot \vec{B}, \qquad (2.32)$$

$$(\vec{r} \times \vec{B})^2 = \vec{B}^2 \vec{r}_\perp^2, \qquad (2.33)$$

---

[1] To evaluate those vector products, the following relations were used:

$$(\vec{A} \times \vec{B}) \cdot (\vec{C} \times \vec{D}) = (\vec{A} \cdot \vec{C})(\vec{B} \cdot \vec{D}) - (\vec{A} \cdot \vec{D})(\vec{B} \cdot \vec{C}), \qquad (2.29)$$

$$\vec{A} \times \vec{B} = -\vec{B} \times \vec{A}, \qquad (2.30)$$

$$\vec{A} \cdot (\vec{B} \times \vec{C}) = (\vec{A} \times \vec{B}) \cdot \vec{C}. \qquad (2.31)$$

where $\vec{L} = \vec{r} \times \vec{p}$ stands for the angular momentum and

$$\vec{r}_\perp^2 = \vec{r}^2 - \frac{(\vec{r} \cdot \vec{B})^2}{\vec{B}^2} \qquad (2.34)$$

is the projection of the $\vec{r}$ vector on the plane normal to the $\vec{B}$ vector, that, in the present case, is the xy plane, since $\vec{B} = B\hat{k}$.

Thus, the final Hamiltonian is

$$\mathcal{H} = \left(\frac{\vec{p}^2}{2m} - e\phi \right) + \left(\frac{\mu_B}{\hbar} \vec{L} \cdot \vec{B} \right) + \left(\frac{e^2}{8m} B^2 \vec{r}_\perp^2 \right), \qquad (2.35)$$

where

$$\mu_B = \frac{e\hbar}{2m} \qquad (2.36)$$

is the Bohr magneton. The first term represents the energy of the free electron under a specific potential; the second term represents the interaction of the orbital momentum, due to precession of the electron around the magnetic field, with its own magnetic field (as will be discussed soon, it is the Zeeman interaction); and, finally, the last term is the origin of the diamagnetism (discussed in detail in Chapter 7). A detailed discussion on these terms will be provided further in this book.

2.2   Quantum-relativistic approach: Dirac equation

This section will provide and discuss the Dirac Hamiltonian that satisfies the equation:

$$\mathcal{H}_D \Psi = \epsilon \Psi. \qquad (2.37)$$

We will first consider a free electron and then an application of both scalar and vectorial potentials.

2.2.1   Free electron

For a free particle, the Dirac Hamiltonian resumes as

$$\mathcal{H}_D = c\vec{\alpha} \cdot \vec{p} + \beta m c^2, \qquad (2.38)$$

where

$$\vec{\alpha} = \begin{pmatrix} 0 & \vec{\sigma} \\ \vec{\sigma} & 0 \end{pmatrix}, \quad \vec{\beta} = \begin{pmatrix} I & 0 \\ 0 & -I \end{pmatrix}. \qquad (2.39)$$

Above, $\vec{\sigma}$ are the Pauli matrices and I is a 2×2 identity matrix.

It is interesting to note the following properties of α_i and β matrices:

$$\alpha_x^2 = \alpha_y^2 = \alpha_z^2 = \beta^2 = I \tag{2.40}$$

and, in addition, those matrices have an anti-commuting property, i.e., $\alpha_x \alpha_y + \alpha_y \alpha_x = 0$.

Let us now express the Dirac Hamiltonian in a matrix form

$$\mathcal{H}_D = \begin{pmatrix} mc^2 & c\vec{\sigma} \cdot \vec{p} \\ c\vec{\sigma} \cdot \vec{p} & -mc^2 \end{pmatrix}. \tag{2.41}$$

Note that it is a reduced form to write that Hamiltonian, because it is in fact a 4×4 matrix. It is easy to obtain the eigenvalues of this matrix, that are:

$$\epsilon_+ = +\sqrt{\vec{p}^2 c^2 + m^2 c^4}, \tag{2.42}$$

$$\epsilon_- = -\sqrt{\vec{p}^2 c^2 + m^2 c^4}. \tag{2.43}$$

Remember, $\vec{\sigma}^2 = I$. Thus, there are two eigenvalues with equal absolute values: one positive and the other negative. This is one of the most important results of the Dirac Hamiltonian, that could predict the existence of *positrons*, those particles with the negative energy (while electrons have the positive energy).

The above equations can be rewritten as (considering only electrons):

$$\epsilon_+ = mc^2 \left(1 + \frac{\vec{p}^2 c^2}{m^2 c^4} \right)^{1/2} \tag{2.44}$$

and therefore, for the low relativistic limit (where $\vec{p}^2 \ll m^2 c^2$), it is possible to write[2]:

$$\epsilon_+ = mc^2 + \frac{\vec{p}^2}{2m} - \frac{\vec{p}^4}{8m^3 c^2}. \tag{2.45}$$

The above energy corresponds therefore to a free electron in the low relativistic limit, where the first term is the rest energy, the second term the classical one, and the third term corresponds to the first relativistic correction of the electron moment $\vec{p}$.

2.2.2 Electron in a scalar potential

For the sake of simplicity, let us continue with our development considering a finite scalar potential $\phi = \phi(\vec{r})$ and zero vector potential $\vec{A} = 0$. It means that the magnetic field is also zero (from Eq. (2.A.7)) and the electric field $\vec{E} = -\vec{\nabla}\phi$ (from Eq. (2.A.11)). We are considering, for instance, an electron under the electric field of a proton without magnetic moment (since there is no magnetic field under the electron, that would be created by the proton magnetic moment).

---

[2]For $x \to 0$ limit, the following approximation is valid:

$$(1 + x)^n \approx 1 + nx + \frac{1}{2}(n - 1)nx^2.$$

Thus, the Hamiltonian reads as

$$\mathcal{H}_D = c\vec{\alpha} \cdot \vec{p} + \beta mc^2 - e\phi. \tag{2.46}$$

Remember, from Eq. (2.21), that an electron under a scalar potential ϕ has an $-e\phi$ contribution to the Hamiltonian. Then, rewriting Eq. (2.37) in a matricial form, we obtain:

$$\begin{pmatrix} \epsilon + e\phi - mc^2 & -c\vec{\sigma} \cdot \vec{p} \\ -c\vec{\sigma} \cdot \vec{p} & \epsilon + e\phi + mc^2 \end{pmatrix} \begin{pmatrix} \psi_1 \\ \psi_2 \end{pmatrix} = 0. \tag{2.47}$$

It is easy to see that we have two coupled equations, and to reach our objective (some specific contributions of the Dirac Hamiltonian), we must decouple these equations. Thus, the relation given below is true:

$$\psi_2 = \frac{c\vec{\sigma} \cdot \vec{p}}{(\epsilon + e\phi + mc^2)} \psi_1 \tag{2.48}$$

and then, it is possible to obtain the Dirac Equation for only ψ_1:

$$(\epsilon + e\phi - mc^2)\psi_1 - c^2(\vec{\sigma} \cdot \vec{p})[\epsilon + e\phi + mc^2]^{-1}(\vec{\sigma} \cdot \vec{p})\psi_1 = 0. \tag{2.49}$$

Considering $\epsilon' = \epsilon - mc^2$ and the limit where $2mc^2 \gg \epsilon' + e\phi$, the above equation can therefore be rewritten as[3]

$$(\epsilon' + e\phi)\psi_1 - \frac{1}{2m}(\vec{\sigma} \cdot \vec{p})\left[1 - \frac{\epsilon' + e\phi}{2mc^2}\right](\vec{\sigma} \cdot \vec{p})\psi_1 = 0. \tag{2.50}$$

Taking advantage of the general identity:

$$(\vec{\sigma} \cdot \vec{A})(\vec{\sigma} \cdot \vec{B}) = \vec{A} \cdot \vec{B} + i\vec{\sigma} \cdot \left(\vec{A} \times \vec{B}\right), \tag{2.51}$$

it is possible to obtain:

$$(\vec{\sigma} \cdot \vec{p})(\vec{\sigma} \cdot \vec{p}) = \vec{p}^2 \tag{2.52}$$

---

[3] For the $x \to 0$ limit, the following approximation is valid:

$$(1 + x)^n \approx 1 + nx + \frac{1}{2}(n - 1)nx^2.$$

and[4]

$$(\vec{\sigma} \cdot \vec{p})e\phi(\vec{\sigma} \cdot \vec{p}) = -i\hbar\{[\vec{\sigma} \cdot \vec{\nabla}(e\phi)](\vec{\sigma} \cdot \vec{p})\}$$
$$= -i\hbar\{\vec{\nabla}(e\phi) \cdot \vec{p} + i\vec{\sigma} \cdot [\vec{\nabla}(e\phi) \times \vec{p}]\}$$
$$= -\hbar^2 \frac{\partial(e\phi)}{\partial r}\frac{\partial}{\partial r} + \frac{2}{r}\frac{\partial(e\phi)}{\partial r}\vec{S} \cdot \vec{L}. \tag{2.55}$$

Now, the Dirac Equation reads as

$$(\epsilon' + e\phi)\psi_1 - \frac{\vec{p}^2}{2m}\psi_1 + \frac{\epsilon'\vec{p}^2}{(2mc)^2}\psi_1$$
$$+ \frac{1}{(2mc)^2}\left[-\hbar^2 \frac{\partial(e\phi)}{\partial r}\frac{\partial}{\partial r} + \frac{2}{r}\frac{\partial(e\phi)}{\partial r}\vec{S} \cdot \vec{L}\right]\psi_1 = 0. \tag{2.56}$$

Considering $\epsilon' = \epsilon - mc^2$ and Eq. (2.45) we have:

$$\epsilon' \approx mc^2 + \frac{\vec{p}^2}{2m} - mc^2 \approx \frac{\vec{p}^2}{2m}. \tag{2.57}$$

Thus, the third term of Eq. (2.56) resumes as

$$\frac{\epsilon'\vec{p}^2}{(2mc)^2} \approx \frac{\vec{p}^4}{8m^3c^2}. \tag{2.58}$$

The final Dirac Equation can be written as

$$\mathcal{H}_D\psi_1 = \epsilon\psi_1, \tag{2.59}$$

where the Dirac Hamiltonian is

$$\mathcal{H}_D = mc^2 + \frac{\vec{p}^2}{2m} - \frac{\vec{p}^4}{8m^3c^2} - e\phi + \frac{\hbar^2}{4m^2c^2}\frac{\partial(e\phi)}{\partial r}\frac{\partial}{\partial r} - \frac{1}{2m^2c^2}\frac{1}{r}\frac{\partial(e\phi)}{\partial r}\vec{S} \cdot \vec{L}. \tag{2.60}$$

Note that in the present case we are working with only the scalar potential and therefore:

$$\vec{E} = -\vec{\nabla}\phi = -\frac{1}{e}\vec{F} = \frac{1}{e}\vec{\nabla}V, \tag{2.61}$$

where V is the real potential in which the electron is. Thus, $V = -e\phi$.

A simple comparison of Eqs. (2.45) and (2.60) leads us to conclude that the scalar potential ϕ rules the spin-orbit $\vec{S} \cdot \vec{L}$ coupling.

[4]In spherical coordinates, the gradient of a scalar field is written as

$$\vec{\nabla}\phi(\vec{r}) = \vec{\nabla}\phi(r, \theta, \varphi) = \frac{\partial}{\partial r}\phi(r, \theta, \varphi)\hat{r} + \frac{1}{r}\frac{\partial}{\partial\theta}\phi(r, \theta, \varphi)\hat{\theta} + \frac{1}{r\sin\theta}\frac{\partial}{\partial\varphi}\phi(r, \theta, \varphi)\hat{\varphi}. \tag{2.53}$$

The present problem has spherical symmetry ($\phi(\vec{r}) = \phi(r)$), and therefore

$$\vec{\nabla}\phi(\vec{r}) = \vec{\nabla}\phi(r) = \frac{\vec{r}}{r}\frac{\partial}{\partial r}\phi(r). \tag{2.54}$$

In addition, to evaluate the calculus corresponding to this footnote, we also need to consider $\vec{L} = \vec{r} \times \vec{p}$, $\vec{p} = -i\hbar\vec{\nabla}$, and $\vec{S} = \frac{\hbar}{2}\vec{\sigma}$ ($\vec{S}$ means the spin of the system and it will be in the discussed next chapter).

2.2.3 Electron in a scalar and vector potential

Let us now assume that the proton (around which the electron is) has a magnetic moment and therefore produces a magnetic field that can be felt by the electron. In this case, $\vec{A} = \vec{A}(\vec{r}) \neq 0$ and then the following transformation must be done (based on Eq. (2.21)):

$$\vec{p} \mapsto \vec{p} + e\vec{A}. \tag{2.62}$$

Thus, the Dirac Hamiltonian reads as

$$\mathcal{H}_D = c\vec{\alpha} \cdot \left(\vec{p} + e\vec{A}\right) + \beta mc^2 - e\phi \tag{2.63}$$

and then Eq. (2.50) can be rewritten as

$$(\epsilon' + e\phi)\psi_1 - \frac{1}{2m}[\vec{\sigma} \cdot (\vec{p} + e\vec{A})]\left[1 - \frac{\epsilon' + e\phi}{2mc^2}\right][\vec{\sigma} \cdot (\vec{p} + e\vec{A})]\psi_1 = 0. \tag{2.64}$$

Let us focus our attention to the corrections of the term

$$\frac{1}{2m}[\vec{\sigma} \cdot (\vec{p} + e\vec{A})][\vec{\sigma} \cdot (\vec{p} + e\vec{A})] \tag{2.65}$$

and discharge those corrections from the term

$$[\vec{\sigma} \cdot (\vec{p} + e\vec{A})]\left[\frac{\epsilon' + e\phi}{2mc^2}\right][\vec{\sigma} \cdot (\vec{p} + e\vec{A})] \tag{2.66}$$

that are quite beyond the scope of the present objective.

From the identity relation presented in Eq. (2.51), it is easy to show that[5]

$$\frac{1}{2m}[\vec{\sigma} \cdot (\vec{p} + e\vec{A})][\vec{\sigma} \cdot (\vec{p} + e\vec{A})] = \frac{(\vec{p} + e\vec{A})^2}{2m} + 2\frac{\mu_B}{\hbar}\vec{S} \cdot \vec{B} \tag{2.67}$$

and then we can replace the second term of Eq. (2.60) by the above result. Thus, the final Dirac Hamiltonian reads as:

$$\mathcal{H}_D = mc^2 + \frac{(\vec{p} + e\vec{A})^2}{2m} + 2\frac{\mu_B}{\hbar}\vec{S} \cdot \vec{B} - \frac{\vec{p}^4}{8m^3c^2} - e\phi + \frac{\hbar^2}{4m^2c^2}\frac{\partial(e\phi)}{\partial r}\frac{\partial}{\partial r}$$
$$- \frac{1}{2m^2c^2}\frac{1}{r}\frac{\partial(e\phi)}{\partial r}\vec{S} \cdot \vec{L}. \tag{2.68}$$

Note that the vector potential is responsible for the Zeeman term (third above).

---

[5]Considering $\vec{B} = \vec{\nabla} \times \vec{A}$, $\vec{S} = \frac{\hbar}{2}\vec{\sigma}$, and $\mu_B = \frac{e\hbar}{2m}$.

2.2.4 Particular case: Uniform magnetic field

Analogously to Section 2.1.1, consider the particular case where the magnetic field points along the z direction, i.e., $\vec{B} = B\hat{k}$. Thus, the second and fifth terms of the previous equation can be replaced by Eq. (2.35). The Dirac Hamiltonian for an electron under a uniform magnetic field $\vec{B} = B\hat{k}$ reads then as:

$$\mathcal{H}_D = mc^2 + \frac{\vec{p}^2}{2m} - \frac{\vec{p}^4}{8m^3c^2} - e\phi + \frac{\mu_B}{\hbar}\left(2\vec{S} + \vec{L}\right)\cdot\vec{B} + \frac{e^2}{8m}B^2\vec{r}_\perp^2$$

$$+ \frac{\hbar^2}{4m^2c^2}\frac{\partial(e\phi)}{\partial r}\frac{\partial}{\partial r} - \frac{1}{2m^2c^2}\frac{1}{r}\frac{\partial(e\phi)}{\partial r}\vec{S}\cdot\vec{L}. \qquad (2.69)$$

The first four terms represent the classical relativistic correction to the Schrödinger equation; the fifth term is the Zeeman interaction; the sixth the diamagnetic term; the seventh the relativistic correction to the potential energy (called as Darwin term), and, finally, the last one is the spin-orbit coupling. These contributions to the Dirac Hamiltonian will be discussed deeper further in this book.

Complements

2.A Maxwell equations and the gauges

The scope of this book does not include a (deep) description of the Maxwell equations, since these contains much information that would deviate our attention from the real scope (magnetic properties of some materials). Due to this reason, only a brief Complement is devoted to this issue, with a simple presentation of that set of equations. Thus, the microscopic formulation is

$$\vec{\nabla}\cdot\vec{E} = \frac{1}{\epsilon_0}\rho, \qquad (2.A.1)$$

$$\vec{\nabla}\times\vec{E} = -\frac{\partial\vec{B}}{\partial t}, \qquad (2.A.2)$$

$$\vec{\nabla}\cdot\vec{B} = 0, \qquad (2.A.3)$$

$$\vec{\nabla}\times\vec{B} = \mu_0\vec{j} + \epsilon_0\mu_0\frac{\partial\vec{E}}{\partial t}, \qquad (2.A.4)$$

where $\rho = \rho(\vec{r}, t)$ is the volume charge density, $\vec{j} = \vec{j}(\vec{r}, t)$ is the volume current density, $\vec{B} = \vec{B}(\vec{r}, t)$ is the magnetic field induction, $\vec{E} = \vec{E}(\vec{r}, t)$ is the electric field, μ_0 and ϵ_0 are the permeability and permittivity of the vacuum, respectively, related (these two last), to the light speed in the vacuum:

$$c^2 = \frac{1}{\mu_0\epsilon_0}. \qquad (2.A.5)$$

From vector analysis we know that the divergence of the rotational of any vector field, e.g., $\vec{A} = \vec{A}(\vec{r}, t)$, is null, i.e.,

$$\vec{\nabla} \cdot (\vec{\nabla} \times \vec{A}) = 0. \tag{2.A.6}$$

From the above relation and considering Eq. (2.A.3), we can write:

$$\vec{B} = \vec{\nabla} \times \vec{A}. \tag{2.A.7}$$

If we substitute the above result into Eq. (2.A.2), we find

$$\vec{\nabla} \times \left(\vec{E} + \frac{\partial \vec{A}}{\partial t} \right) = 0. \tag{2.A.8}$$

However, also from vector analysis, we know that the rotational of the gradient of any scalar field, e.g., $\phi = \phi(\vec{r}, t)$, is null, i.e.,

$$\vec{\nabla} \times (\vec{\nabla} \phi) = 0. \tag{2.A.9}$$

From the above relations it is possible to define

$$-\vec{\nabla} \phi = \vec{E} + \frac{\partial \vec{A}}{\partial t}, \tag{2.A.10}$$

where the negative signal was introduced due to future convenience.

These quantities derived above are known as *vector potential* $\vec{A}(\vec{r}, t)$ and *scalar potential* $\phi(\vec{r}, t)$. From these potentials, the magnetic field $\vec{B}(\vec{r}, t)$ and electric field $\vec{E}(\vec{r}, t)$ are therefore determined:

$$\vec{E} = -\vec{\nabla} \phi - \frac{\partial \vec{A}}{\partial t}, \tag{2.A.11}$$

$$\vec{B} = \vec{\nabla} \times \vec{A}. \tag{2.A.12}$$

From the discussion above, it is easy to see that a certain magnetic field and electric field $\{\vec{B}, \vec{E}\}$ can be described by infinity possibilities of potentials, also known as *gauges*, $\{\vec{A}, \phi\}$. These potentials that create the same electric and magnetic fields are known as *equivalent gauges*. If we known a gauge $\{\vec{A}, \phi\}$ that produces a field $\{\vec{B}, \vec{E}\}$, all of the equivalent gauges $\{\vec{A}', \phi'\}$ can be written as:

$$\vec{A}'(\vec{r}, t) = \vec{A}(\vec{r}, t) + \vec{\nabla} f(\vec{r}, t), \tag{2.A.13}$$

$$\phi'(\vec{r}, t) = \phi(\vec{r}, t) - \frac{\partial}{\partial t} f(\vec{r}, t), \tag{2.A.14}$$

where $f(\vec{r}, t)$ is any scalar field. It is easy to prove that equivalent gauges $\{\vec{A}, \phi\}$ and $\{\vec{A}', \phi'\}$ can create the same magnetic field and electric field $\{\vec{B}, \vec{E}\}$. This proof is left as an exercise to the reader.

3 Angular Momenta

The orbital moment $\vec{L}$ has its origin on the precessing movement of the electron around the nucleus. It is found in both, classical and quantum cases and is easy to understand. On the other hand, the spin momentum $\vec{S}$ is only obtained from the Dirac equation (the relativistic version of the Schrödinger equation, described in detail in the previous chapter) and has no classic analogy. Figure 3.1 depicts this scenario. These angular momenta (orbital and spin) must be added and define therefore the total angular momentum $\vec{J} = \vec{L} + \vec{S}$. See Figure 3.2.

3.1 Angular momentum algebra

All of the angular momenta (spin, orbital, and total) obey the same algebra here described and then, instead of J (as will be used here), it is possible to substitute that by either L or S. Thus, let us first consider the angular momentum in Cartesian coordinates:

$$\vec{J} = J_x \hat{i} + J_y \hat{j} + J_z \hat{k}. \tag{3.1}$$

These components obey the following commutation relation:

$$[J_x, J_y] = i\hbar J_z, \tag{3.2}$$
$$[J_y, J_z] = i\hbar J_x, \tag{3.3}$$
$$[J_z, J_x] = i\hbar J_y. \tag{3.4}$$

The reason in which all of the angular momentum algebra is valid for all of the angular momenta came from the above set of equations, i.e., spin, orbital, and total angular momenta follow the same commutation relation.

Figure 3.1 Orbital and spin momenta. Classical picture.

Fundamentals of Magnetism. http://dx.doi.org/10.1016/B978-0-12-405545-2.00003-5

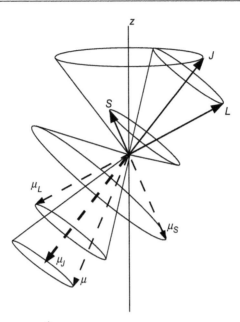

Figure 3.2 Orbital $\vec{L}$ and spin $\vec{S}$ angular momenta added to obtain the total angular momentum $\vec{J}$. These vectors are $\sqrt{u(u+1)}\hbar$ in size, where u can be l, s, or j. The associated magnetic moments are also shown, in accordance with the ideas described further in this chapter. Note the total magnetic moment $\vec{\mu}$ has two components, one parallel and other normal to $\widehat{J}$; and only μ_J (parallel to $\widehat{J}$) contributes to the magnetic energy.

Note these components do not commute each other and therefore these cannot be diagonalized simultaneously. Thus, we need to consider other quantity: the square of the total angular momentum operator:

$$J^2 = J_x J_x + J_y J_y + J_z J_z. \tag{3.5}$$

This new operator commutes with each of those components:

$$[J^2, J_u] = 0 \ (u = x, y, z) \tag{3.6}$$

and therefore these can be diagonalized simultaneously. We can now define the basis $|j, m_j\rangle$, where J^2 and J_z are diagonals (this last was chosen by convenience; but we could choose either J_x or J_y). Thus,

$$J^2|j, m_j\rangle = j(j+1)\hbar^2|j, m_j\rangle, \tag{3.7}$$

where j is the total angular momentum of the system, and

$$J_z|j, m_j\rangle = m_j \hbar|j, m_j\rangle, \tag{3.8}$$

where $m_j = -j, -j+1, \ldots, +j-1, +j$, is the projection of the total angular momentum $\vec{J}$ on the z-axis.

We have thus a basis where both, J^2 and J_z are diagonals. However, we do not know all of the total moment, since the components J_x and J_y remain to be determined. To go further we need to define two operators, named ladder operators:

$$J_+ = J_x + iJ_y \quad \text{and} \quad J_- = J_x - iJ_y \tag{3.9}$$

that commutes as

$$[J_+, J_-] = 2\hbar J_z, \quad [J_z, J_\pm] = \pm\hbar J_\pm, \quad [J^2, J_\pm] = 0. \tag{3.10}$$

Thus, it is possible to determine:

$$J_\pm|j, m_j\rangle = \sqrt{(j \mp m_j)(j \pm m_j + 1)}\hbar|j, m_j \pm 1\rangle . \tag{3.11}$$

From Eq. (3.9), it is possible to write:

$$J_x = \frac{1}{2}(J_+ + J_-) \quad \text{and} \quad J_y = \frac{1}{2i}(J_+ - J_-) \tag{3.12}$$

and therefore determine the final operator $\vec{J}$.

Few words about the vectorial space in which these vectors are: the Hilbert space. Mathematically speaking, it is a generalization of the Euclidian space. For a total angular momentum J, there are $2j + 1$ different states (m_j possibilities). Thus, the matrices obtained from above have dimension $(2j + 1) \times (2j + 1)$. We leave as an exercise to write the matrix operators J_x, J_y, and J_z, based on the above description.

3.2 Addition of angular momenta

The question discussed in this section is how to couple the orbital and spin momenta. First, let us consider the total angular moment:

$$\vec{J} = \vec{L} + \vec{S}, \tag{3.13}$$

where each angular momentum (orbital and spin) has its own subspaces. Thus, we can write:

$$S_z|s, m_s\rangle = m_s\hbar|s, m_s\rangle, \tag{3.14}$$
$$S^2|s, m_s\rangle = s(s + 1)\hbar^2|s, m_s\rangle$$

and

$$L_z|l, m_l\rangle = m_l\hbar|l, m_l\rangle, \tag{3.15}$$
$$L^2|l, m_l\rangle = l(l + 1)\hbar^2|l, m_l\rangle.$$

The total space is described by the set $\{S^2, L^2, S_z, L_z\}$, represented by the state $|s, l, m_s, m_l\rangle$. The quantities $\{S^2, L^2, J^2, J_z\}$ commute each other and then can also be diagonalized simultaneously. Thus, for this new basis $|s, l, j, m_j\rangle$ the operators J_z and J^2 are diagonals:

$$J_z|s, l, j, m_j\rangle = m_j \hbar |s, l, j, m_j\rangle, \tag{3.16}$$
$$J^2|s, l, j, m_j\rangle = j(j+1)\hbar^2 |s, l, j, m_j\rangle.$$

Above, the quantum numbers j and m_j must satisfy the conditions:

$$|s - l| \leq j \leq s + l \tag{3.17}$$

and

$$-j \leq m_j \leq +j. \tag{3.18}$$

The above result is the main one of this section; but, for the sake of completeness, let us describe the matrix that allows to change the system basis. For this purpose, let us consider the following closure relation:

$$\sum_{m_s} \sum_{m_l} |s, l, m_s, m_l\rangle\langle s, l, m_s, m_l| = \mathbb{I}, \tag{3.19}$$

where $\mathbb{I}$ is the identity operator. If we multiply the above relation on the right-hand side of $|s, l, j, m_j\rangle = |s, l, j, m_j\rangle$, we obtain:

$$|s, l, j, m_j\rangle = \sum_{m_s} \sum_{m_l} \langle s, l, m_s, m_l | s, l, j, m_j\rangle |s, l, m_s, m_l\rangle, \tag{3.20}$$

where

$$\langle s, l, m_s, m_l | s, l, j, m_j\rangle = (-1)^{s-l+m_j}(2j+1)^{1/2} \begin{pmatrix} s & l & j \\ m_s & m_l & -m_j \end{pmatrix} \tag{3.21}$$

are the well-known Clebsch–Gordan coefficients, that, on its turn, can be written in terms of the Wigner $3j$-symbols (the 2×3 symbol on the right). These coefficients are always real and constitute a matrix with the dimension of the total Hilbert space, that is $(2s + 1)(2l + 1)$. To take the Clebsch–Gordan coefficients, we need to solve the Wigner $3j$-symbol that, on its turn, can be evaluated from the Racah formula:

$$\begin{pmatrix} a & b & c \\ A & B & C \end{pmatrix} = (-1)^{a-b-C} Q^{1/2}(\Delta_{abc})^{1/2} \sum_{n=n_{min}}^{n_{max}} \frac{(-1)^n}{f(n)}, \tag{3.22}$$

where

$$Q = (a + A)!(a - A)!(b + B)!(b - B)!(c + C)!(c - C)!, \tag{3.23}$$

$$\Delta_{abc} = \frac{(a + b - c)!(b + c - a)!(c + a - b)!}{(a + b + c + 1)!}, \tag{3.24}$$

$$f(n) = n!(n - n_1)!(n - n_2)!(n_3 - n)!(n_4 - n)!(n_5 - n)!, \tag{3.25}$$

$$n_1 = -c + b - A, \tag{3.26}$$

$$n_2 = -c + a + B, \tag{3.27}$$

$$n_3 = a + b - c,$$

$$n_4 = b + B,$$

$$n_5 = a - A,$$

$$n_{\min} = \max\{0; \max\{n_1, n_2\}\}, \tag{3.28}$$

and, finally,

$$n_{\max} = \min\{n_3; \min\{n_4, n_5\}\}. \tag{3.29}$$

The Racah formula is zero for the case in which the conditions below are not satisfied:

$$A + B + C = 0, \tag{3.30}$$

$$|a - b| \le c \le a + b \tag{3.31}$$

and, consequently, the Clebsch–Gordan coefficients are also zero for the case in which these conditions are not satisfied:

$$m_j = m_s + m_l , \tag{3.32}$$

$$|s - l| \le j \le s + l. \tag{3.33}$$

3.3   Magnetic moment

To discuss the magnetic moments, let us start from the Zeeman Hamiltonian (fifth term of Eq. (2.69)), i.e., the interaction of both, spin and orbital momenta with a magnetic field:

$$\mathcal{H}_{S,L} = \frac{\mu_B}{\hbar} \left(2\vec{S} + \vec{L} \right) \cdot \vec{B}. \tag{3.34}$$

First, let us focus our attention to the spin contribution:

$$\mathcal{H}_{S,L}^{(S)} = 2\frac{\mu_B}{\hbar}\vec{S} \cdot \vec{B} \tag{3.35}$$

and rewrite this as:

$$\mathcal{H}_{S,L}^{(S)} = -\vec{\mu}_S \cdot \vec{B}, \tag{3.36}$$

where

$$\vec{\mu}_S = -g_S \frac{\mu_B}{\hbar} \vec{S} = \gamma_S \vec{S}. \tag{3.37}$$

Above, $g_S = 2$ and $\gamma_S = -e/m$ is the *spin* gyromagnetic factor.

Note we defined (but it can be formally shown, as will be done further in this book), a negative magnetic energy in terms of the magnetic moment (Eq. (3.36)). It is reasonable, since the minimum energy is the magnetic moment aligned with the magnetic field.

In analogy to the spin case, we can also deal with only the orbital contribution:

$$\mathcal{H}_{S,L}^{(L)} = \frac{\mu_B}{\hbar} \vec{L} \cdot \vec{B} \tag{3.38}$$

that can be rewritten as:

$$\mathcal{H}_{S,L}^{(L)} = -\vec{\mu}_L \cdot \vec{B}, \tag{3.39}$$

where

$$\vec{\mu}_L = -g_L \frac{\mu_B}{\hbar} \vec{L} = \gamma_L \vec{L}. \tag{3.40}$$

It is easy to see that $g_L = 1$ and $\gamma_L = -e/2m$; this last is the *orbital* gyromagnetic factor.

Thus, analogously to Eqs. (3.36) and (3.39), we can write:

$$\mathcal{H}_{S,L} = -\vec{\mu} \cdot \vec{B}. \tag{3.41}$$

However, we know that

$$\mathcal{H}_{S,L} = \mathcal{H}_{S,L}^{(L)} + \mathcal{H}_{S,L}^{(S)} \tag{3.42}$$

and therefore

$$\begin{aligned} \vec{\mu} &= \vec{\mu}_L + \vec{\mu}_S, \\ &= -\frac{\mu_B}{\hbar} \left(g_L \vec{L} + g_S \vec{S} \right). \end{aligned} \tag{3.43}$$

See Figure 3.2 to understand this scenario. The total magnetic moment $\vec{\mu}$ is not parallel to $\vec{J}$, due to the different gyromagnetic factors of the orbital and spin momenta. However, $\vec{\mu}$ precesses around the $\vec{J}$ direction and, as a consequence, the component of $\vec{\mu}$ normal to the total angular momentum direction is zero on time average and, therefore, only μ_J (antiparallel to $\vec{J}$) contributes to the magnetic energy.

If we multiply both sides of the above equation by

$$\hat{J} = \frac{\vec{J}}{J}, \tag{3.44}$$

we obtain:

$$\vec{\mu} \cdot \hat{J} = \mu_J = -\frac{\mu_B}{\hbar} \left(\frac{\vec{L} \cdot \vec{J}}{J} + 2 \frac{\vec{S} \cdot \vec{J}}{J} \right). \tag{3.45}$$

We know that $\vec{J} = \vec{L} + \vec{S}$ and then:

$$\vec{S}^2 = (\vec{J} - \vec{L}) \cdot (\vec{J} - \vec{L}) = \vec{J}^2 + \vec{L}^2 - 2\vec{L} \cdot \vec{J}, \tag{3.46}$$
$$\vec{L}^2 = (\vec{J} - \vec{S}) \cdot (\vec{J} - \vec{S}) = \vec{J}^2 + \vec{S}^2 - 2\vec{S} \cdot \vec{J}, \tag{3.47}$$

that can be inserted into Eq. (3.45) to obtain:

$$\mu_J = -\frac{\mu_B}{2J\hbar} \left(3\vec{J}^2 - \vec{L}^2 + \vec{S}^2 \right). \tag{3.48}$$

Thus, considering

$$\vec{\mu}_J = -g_J \frac{\mu_B}{\hbar} \vec{J} \tag{3.49}$$

and comparing with Eq. (3.48), we can obtain:

$$g_J = \frac{3\vec{J}^2 - \vec{L}^2 + \vec{S}^2}{2\vec{J}^2}. \tag{3.50}$$

These square angular momenta are operators and the eigenvalues are known (see Section 3.1). Thus, after a minor algebraic change we find:

$$g \equiv g_J = 1 + \frac{j(j+1) - l(l+1) + s(s+1)}{2j(j+1)}, \tag{3.51}$$

where g is known as Landé factor.

Finally, the Zeeman Hamiltonian can be written as

$$\mathcal{H}_{S,L} = -\vec{\mu}_J \cdot \vec{B}, \tag{3.52}$$

where $\vec{\mu}_J$ is given by Eq. (3.49). The eigenvalues (considering $\vec{B} = B\hat{k}$) are

$$\varepsilon_{m_j} = g m_j \mu_B B. \tag{3.53}$$

Above, the eigenvalues of J_z are $m_j \hbar$ (see Section 3.1 for further details).

3.4 Angular momenta of atoms

The ideas described above are for one electron around one proton; a hypothetical situation to be applied to a real system, but, in any case, gives us the fundamental idea behind what occurs in an atom. As known, atoms have a lot of electrons, with complete shells (that do not contribute to the effective angular momenta) and an incomplete shell (that does contribute to the effective angular momenta). For light atoms ($Z < 75$), the spin momentum of each electron couples to produce an effective spin momentum:

$$\vec{S} = \sum_i \vec{S}_i, \tag{3.54}$$

i.e., the sum of all of the spin momenta (of each electron). Analogously, the orbital momenta are also coupled and then the effective orbital momentum reads as

$$\vec{L} = \sum_i \vec{L}_i. \tag{3.55}$$

Then, these effective momenta interact through the spin-orbit coupling (last term of Eq. (2.69)):

$$\mathcal{H}_{S\mathcal{L}} = \zeta \vec{S} \cdot \vec{L}, \tag{3.56}$$

where ζ is the spin-orbit parameter and $\vec{S}$ and $\vec{L}$ are given by Eqs. (3.54) and (3.55); analogously to the case of one electron and one proton. This book will be focused on this case: light atoms ruled by the spin-orbit coupling.

For the sake of completeness, for heavy atoms ($Z > 75$), each electron couples its $\vec{S}$ and $\vec{L}$ to produce $\vec{J}$ and then, the total angular momentum of each electron couples to produce the effective total angular momentum:

$$\vec{J} = \sum_i \vec{J}_i. \tag{3.57}$$

This case is ruled by the *J-J* coupling.

Note that above we used for both, individual and effective momenta, the same letter and typeset. It is because both are ruled by the same angular momentum algebra.

Let us go further on the discussion of light atoms. The values of the effective $\vec{S}$ and $\vec{L}$ are given by the Hund's rules:

First: From the Pauli exclusion principle, only two electrons can be in the same spatial orbit (one with spin up and the other with spin down). To minimize the electronic repulsion, the orbitals are filled one-by-one and then, after all of them occupied, the first orbital receives the second electron (with spin down). It means that the effective spin $\vec{S}$ must be maximized; and this is the first Hund rule.

Second: For a given $\vec{S}$, there are a plenty of possible effective orbital $\vec{L}$ values. The point is: which one minimizes the energy? The response is also behind the electronic repulsion. If all of the electrons precess around the nucleus in the same sense, it minimizes the chance of one electron to meet its neighbor. Thus, the value to maximize $\vec{L}$ then minimize the energy: it is the second Hund rule.

Third: This last rule is not intuitive as those mentioned above. It has the origin on the spin-orbit coupling:

$$\mathcal{H}_{S\mathcal{L}} = \zeta \vec{S} \cdot \vec{L}. \tag{3.58}$$

This parameter is positive when the shell is less than half-filled; and negative when the shell is more than half-filled. Thus, for the first case, $\vec{S}$ must be antiparallel to $\vec{L}$ to minimize the spin-orbit coupling; and then $\vec{J} = \vec{L} - \vec{S}$ is the fundamental state. For the second case, $\vec{S}$ must be parallel to $\vec{L}$ and then $\vec{J} = \vec{L} + \vec{S}$ is the fundamental state.

Table 3.1 Diagram to fill Sm^{3+} orbitals. Numbers mean the orbital quantum number m_l, while each arrow means one electron, with a defined spin (either up or down).

+3	+2	+1	0	-1	-2	-3
↑	↑	↑	↑	↑		

Let us consider a simple case: Sm^{3+} (Samarium is a rare earth, and this kind of element will be studied in detail further in this book). The electronic configuration of this element is

$$Sm : [Xe]4f^6 6s^2 \tag{3.59}$$

and then

$$Sm^{3+} : [Xe]4f^5. \tag{3.60}$$

Let us recover the well-known diagram to fill orbitals (see Table 3.1). From the first Hund rule, we know that the effective spin momentum is the simple sum of the individual spins. Then, $s = 1/2+1/2+1/2+1/2+1/2 = 5/2$ (note $\vec{S}$ is a vector, S is a quantum operator, and s is a quantum number, that measures S, i.e., $S^2|\cdots\rangle = s(s+1)\hbar^2|\cdots\rangle$). From the second Hund rule, and analogously to before, $l = +3+2+1+0-1 = +5$. Finally, from the third Hund rule, we know that this case has the shell less than half filled, and then the ground multiplet is $j_0 = |s - l| = 5/2$.

Tables 3.2 and 3.3 and Figure 3.3 present the complete diagram and information about the angular momentum configuration for rare earths and transition metals. It is important to stress, but this point will be discussed further in this book, that for rare-earths, those moments are the same, for bounded and free ions, since the 4f shell is shielded by other completed shells. On the contrary, transition metals have those values of orbital moment only for free ions; while bounded atoms change their orbital momentum (and consequently, their total momentum), due to several reasons (like crystal field), better discussed further in this book.

These angular momenta can be written following the Russell-Saunders Term Symbol:

$$^{2s+1}X_j, \tag{3.61}$$

where X is the effective orbital quantum number l, written in terms of the spectroscopic notation (see Table 3.4 to remember). Thus, for the Sm^{3+} case, it is: $^6H_{5/2}$. Note this is the fundamental state. For the case in which $s + l$ is the ground state, the first excited state is $s + l - 1$. Obviously, for the case in which the ground state is $|s - l|$, the first excited state is $|s - l| + 1$.

Table 3.2 Angular moment configuration for $4f^n$ ions. Values for the energy gap Δ between the ground state (GS) and the first excited state (FES), obtained from [1]. *Note:* j_0 and g_0 are, respectively, the total angular momentum and Landé factor of the ground state.

n	R^{3+}	f^n Shell Orbital Quantum Numbers							s	l	j_0	g_0	$g_0 j_0$	GS	FES	Δ (K)
		+3	+2	+1	0	-1	-2	-3								
1	Ce^{3+}	↑							1/2	3	5/2	6/7	15/7	$^2F_{5/2}$	$^2F_{7/2}$	3150
2	Pr^{3+}	↑	↑						1	5	4	4/5	16/5	$^3H_4$	$^3H_5$	3100
3	Nd^{3+}	↑	↑	↑					3/2	6	9/2	8/11	36/11	$^4I_{9/2}$	$^4I_{11/2}$	2750
4	Pm^{3+}	↑	↑	↑	↑				2	6	4	3/5	12/5	$^5I_4$	$^5I_5$	2300
5	Sm^{3+}	↑	↑	↑	↑	↑			5/2	5	5/2	2/7	5/7	$^6H_{5/2}$	$^6H_{7/2}$	1450
6	Eu^{3+} (Sm^{2+})	↑	↑	↑	↑	↑	↑		3	3	0	5^a	0	$^7F_0$	$^7F_1$	500
7	Gd^{3+} (Eu^{2+})	↑	↑	↑	↑	↑	↑	↑	7/2	0	7/2	2	7	$^8S_{7/2}$	$^6P_{7/2}$	43,200
8	Tb^{3+}	↑↓	↑	↑	↑	↑	↑	↑	3	3	6	3/2	9	$^7F_6$	$^7F_5$	2900
9	Dy^{3+}	↑↓	↑↓	↑	↑	↑	↑	↑	5/2	5	15/2	4/3	10	$^6H_{15/2}$	$^6H_{13/2}$	4750
10	Ho^{3+}	↑↓	↑↓	↑↓	↑	↑	↑	↑	2	6	8	5/4	10	$^5I_8$	$^5I_7$	7500
11	Er^{3+}	↑↓	↑↓	↑↓	↑↓	↑	↑	↑	3/2	6	15/2	6/5	9	$^4I_{15/2}$	$^4I_{13/2}$	9350
12	Tm^{3+}	↑↓	↑↓	↑↓	↑↓	↑↓	↑	↑	1	5	6	7/6	7	$^3H_6$	$^3H_5$	11,950
13	Yb^{3+}	↑↓	↑↓	↑↓	↑↓	↑↓	↑↓	↑	1/2	3	7/2	8/7	4	$^2F_{7/2}$	$^2F_{5/2}$	14,800

^{a} See Exercise 6 of Chapter 8. It is easy to show that $g_0 = 2 + l = 2 + s = 5$.

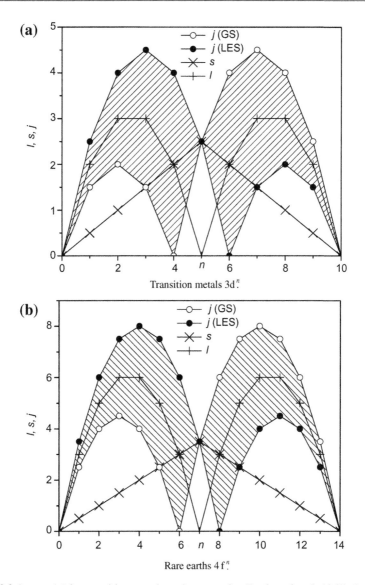

Figure 3.3 l, s, and j for transition metals and rare-earths. For less than half filled shell, the ground state (GS) is $j = |s - l|$ and the last excited state (LES) is $j = s + l$. On the other hand, for more than half filled shell, the ground state is $j = s + l$ and the last excited state is $j = |s - l|$.

Table 3.3 Angular moment configuration for $3d^n$ ions. *Note:* j_0 and g_0 are, respectively, the total angular momentum and Landé factor of the ground state (GS).

		d^n Shell Orbital Quantum Numbers									
n	Metal	+2	+1	0	−1	−2	s	l	j_0	g_0	$g_0 j_0$	GS
1	Ti^{3+}, V^{4+}	↑					1/2	2	3/2	4/5	6/5	$^2D_{3/2}$
2	V^{3+}, Cr^{4+}	↑	↑				1	3	2	2/3	4/3	$^3F_2$
3	Cr^{3+}, Mn^{4+}	↑	↑	↑			3/2	3	3/2	2/5	3/5	$^4F_{3/2}$
4	Cr^{2+}, Mn^{3+}	↑	↑	↑	↑		2	2	0	4^a	0	$^5D_0$
5	Mn^{2+}, Fe^{3+}	↑	↑	↑	↑	↑	5/2	0	5/2	2	5	$^6S_{5/2}$
6	Fe^{2+}, Co^{3+}	↑↓	↑	↑	↑	↑	2	2	4	3/2	6	$^5D_4$
7	Co^{2+}, Ni^{3+}	↑↓	↑↓	↑	↑	↑	3/2	3	9/2	4/3	6	$^4F_{9/2}$
8	Ni^{2+}, Co^+	↑↓	↑↓	↑↓	↑	↑	1	3	4	5/4	5	$^3F_4$
9	Cu^{2+}, Ni^+	↑↓	↑↓	↑↓	↑↓	↑	1/2	2	5/2	6/5	3	$^2D_{5/2}$

^{a} See Exercise 6 of Chapter 8. It is easy to show that $g_0 = 2 + l = 2 + s = 5$.

Table 3.4 Term Symbols use the spectroscopic notation, quite well established. The lowest values of l are related to some specific lines: S-Sharp, P-Principal, D-Diffuse, F-Fundamental, and the remaining values of l follow the alphabetic order, i.e., G, H, I, and so on.

l	0	1	2	3	4	5	6	$\cdots$
X	S	P	D	F	G	H	I	$\cdots$

Complements

3.A   Hydrogen-like atoms

The hydrogen atom, that has an analytic solution when isolated, is an excellent starting point to see the orbitals of each state. The solution (eigenvalues and eigenstates), of this problem starts from the Schrödinger equation and here we will only present the main results, those important to this book.

The wave functions[1] are

$$\Psi_{nlm}(r, \theta, \phi) = R_{nl}(r)Y_l^m(\theta, \phi), \tag{3.A.1}$$

where $Y_l^m(\theta, \phi)$ are the spherical harmonics and

$$R_{nl}(r) = \left(\frac{2}{na_0}\right)^{3/2}\left[\frac{(n-l-1)!}{2n[(n+l)!]}\right]^{1/2} e^{-\rho/2}\rho^l L_{n-l-1}^{2l+1}(\rho). \tag{3.A.2}$$

[1]Represented as $|nlm\rangle$ on the Dirac notation.

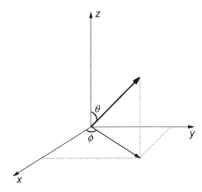

Figure 3.4 Angles and axis chosen to describe the hydrogen atom.

Above, $L_b^a(c)$ stands for the generalized Laguerre polynomials and

$$\rho = \frac{2r}{na_0}, \tag{3.A.3}$$

where $a_0 = 0.529$ Å, is the Bohr radius. Note, the zenith angle θ is defined as that between the vector $\vec{r}$ and the z Cartesian axis, while the azimuth angle ϕ is that between the projection of the $\vec{r}$ vector on the xy plane and the x-axis, as clarified by Figure 3.4. In addition, the wave functions have three quantum numbers:

$$
\begin{array}{ll}
n = 1, 2, 3, \ldots & \text{(principal)} \\
l = 0, 1, 2, 3, \ldots, n-1 & \text{(orbital)} \\
m = -l, -l+1, \ldots, +l-1, +l & \text{(magnetic)}
\end{array}
$$

It is possible to write the radial part (up to $n = 3$):

$$R_{10} = 2\left(\frac{1}{a_0}\right)^{3/2} e^{-r/a_0}, \tag{3.A.4}$$

$$R_{20} = \left(\frac{1}{2a_0}\right)^{3/2} \left(2 - \frac{r}{a_0}\right) e^{-r/2a_0}, \tag{3.A.5}$$

$$R_{21} = \left(\frac{1}{2a_0}\right)^{3/2} \frac{1}{\sqrt{3}} \frac{r}{a_0} e^{-r/2a_0}, \tag{3.A.6}$$

$$R_{30} = 2\left(\frac{1}{3a_0}\right)^{3/2} \left[1 - \frac{2}{3}\frac{r}{a_0} + \frac{2}{27}\left(\frac{r}{a_0}\right)^2\right] e^{-r/3a_0}, \tag{3.A.7}$$

$$R_{31} = \left(\frac{1}{3a_0}\right)^{3/2} \frac{4\sqrt{2}}{3} \left(1 - \frac{1}{6}\frac{r}{a_0}\right) \frac{r}{a_0} e^{-r/3a_0}, \tag{3.A.8}$$

$$R_{32} = \left(\frac{1}{3a_0}\right)^{3/2} \frac{2\sqrt{2}}{27\sqrt{5}} \left(\frac{r}{a_0}\right)^2 e^{-r/3a_0} \tag{3.A.9}$$

and also the angular contribution:

$$Y_0^0 = \frac{1}{2\sqrt{\pi}}, \tag{3.A.10}$$

$$Y_1^0 = \frac{1}{2}\sqrt{\frac{3}{\pi}}\cos\theta, \tag{3.A.11}$$

$$Y_1^{\pm 1} = \frac{1}{2}\sqrt{\frac{3}{2\pi}}\sin\theta e^{\pm i\phi}, \tag{3.A.12}$$

$$Y_2^0 = \frac{1}{4}\sqrt{\frac{5}{\pi}}(3\cos^2\theta - 1), \tag{3.A.13}$$

$$Y_2^{\pm 1} = \frac{1}{2}\sqrt{\frac{15}{2\pi}}\sin\theta\cos\theta e^{\pm i\phi}, \tag{3.A.14}$$

$$Y_2^{\pm 2} = \frac{1}{4}\sqrt{\frac{15}{2\pi}}\sin^2\theta e^{\pm 2i\phi}. \tag{3.A.15}$$

These eigenstates (wave functions) are also important to obtain the infinitesimal probability dP to find the electron in an elemental volume $d\Omega$:

$$dP = |\Psi_{nlm}(r, \theta, \phi)|^2 d\Omega, \tag{3.A.16}$$

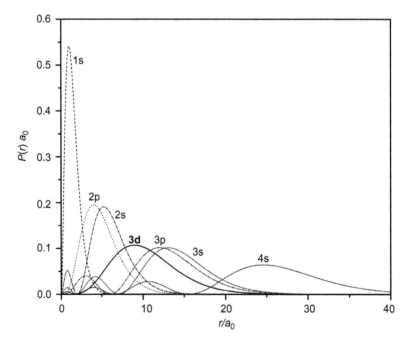

Figure 3.5 Radial distribution probability for shells (following the Madelung energy ordering), up to 3d.

where

$$d\Omega = r^2 \sin(\theta) dr\, d\theta\, d\phi. \tag{3.A.17}$$

Thus:

$$P(r)dr = r^2 |R_{nl}(r)|^2 dr \int_0^\pi \int_0^{2\pi} |Y_l^m(\theta, \phi)|^2 \sin(\theta)\, d\phi\, d\theta. \tag{3.A.18}$$

The integral on the right-hand term has a unitary value, since the spherical harmonics are already normalized. Then we obtain for the radial distribution:

$$P(r) = r^2 |R_{nl}(r)|^2. \tag{3.A.19}$$

These distributions are presented in Figure 3.5, for states up to 3d (following the Madelung energy ordering—see below), and in Figure 3.6, for states from 3d up to 4f.

These figures are for the hydrogen atom, but are important starting points to consider a many-electrons atom. In this sense, note that for 3d metals, like Copper, Iron, Cobalt, and other atoms, the most external subshell is 4s and then, those electrons are used for chemical bounds, as well as some 3d electrons (following the Madelung energy ordering, these two subshells are the most energetic). As a consequence, since these kinds of atoms share the most external subshell to the chemical bounds, the 3d subshell

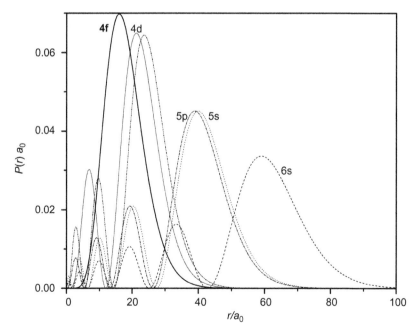

Figure 3.6 Radial distribution probability for shells (following the Madelung energy ordering), up to 4f.

becomes one of the most external and then the unpaired electrons (those responsible for the magnetic properties) have a high mobility and are treated as a fermions gas.

On the other hand, from Figure 3.6, it is easy to see that 4f subshell is an inner one (from the spatial point of view, in spite of being the most energetic). Thus, rare-earth atoms (like Gadolinium, Praseodymium, and other atoms), lose 6s and 4f electrons for the chemical bounds, and the remaining 4f electrons (responsible for the magnetic properties, since those are unpaired) are shielded by the completed outer subshells.

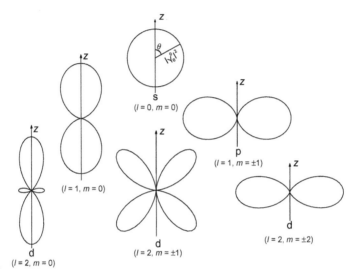

Figure 3.7 Angular dependency of subshells s, p, and d. The distance from the origin to the curve represents the modulus squared of the spherical harmonic associated to the orbital. Due to the independence of ϕ on the modulus squared, a 3D picture is possible from a revolution of these curves around the z-axis.

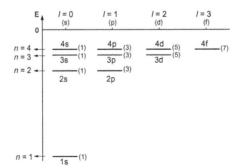

Figure 3.8 Energy spectra for the hydrogen atom. The values in parentheses, near the levels, are the corresponding degenerecy.

Thus, the magnetic behavior of these materials can be described considering localized magnetic moment, with well-defined spin and angular momenta.

Due to the normalization of the spherical harmonics, we can write the angular distribution as:

$$P(\theta, \phi) = |Y_l^m(\theta, \phi)|^2. \tag{3.A.20}$$

It is easy to see that there is no azimuthal ϕ dependence, since this angle only appears into the argument of complex exponentials and then disappears when we take the modulus squared. Thus, the orbitals have only θ dependency. These orbitals are represented in Figure 3.7, that shows the polar plots (not Cartesians), of the modulus squared of the eigenstate of interest. Note that due to this ϕ independence, a 3D view of the orbitals is obtained from a revolution of those curves around the z-axis.

Also note the spherical symmetry of the s state ($l = 0$). To the following subshells $l > 0$ (that has $2l+1$ orbitals), the orbitals are along the z-axis for $m = 0$, and increasing the value of m, it tends to be orthogonal to the z-axis. The angular dependency of these eigenstates is the same for any n shell.

Finally, in what concerns the energy spectra, this can also be obtained from the Schrödinger equation. These values are:

$$E_n = -\frac{E_0}{n^2}, \tag{3.A.21}$$

where $E_0 = -13.6$ eV is the ionization energy of the hydrogen atom. From the energy spectra and the relationship between those three quantum numbers, it is possible to draw schematically those levels, as in Figure 3.8. Note each shell n has degeneracy n^2, i.e., there are n^2 states with the same energy.

These results are therefore for an isolated hydrogen atom. However, what does it occur for a many-electrons atom? There is a change in the energy spectra, following the Madelung energy ordering. The *aufbau* principle (that means, in German, "building

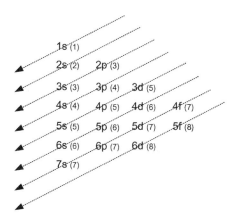

Figure 3.9 Madelung energy ordering: the *aufbau* principle. Between parentheses are the $n + l$ value of each state. See text for further details.

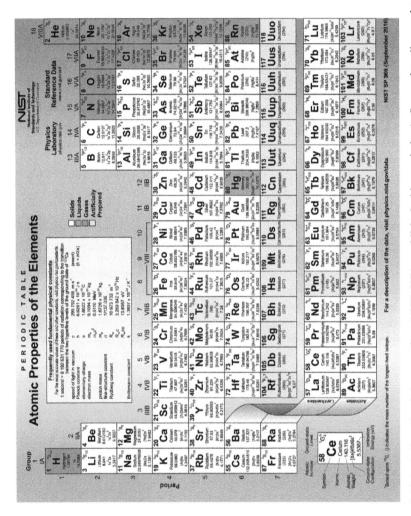

Figure 3.10 Periodic table of elements (from NIST, USA - published with permission.). *Source:* http://www.nist.gov.

up") is based on the hypothesis that atoms are made by adding electrons. The new electron added is then inserted in the most stable orbital available. This energetically favorable state is then given by the Madelung rule, where orbitals with lower $n + l$ values are filled first. If two different orbitals have the same $n + l$ value, then the one with lower n is filled first. See Figure 3.9 for a view of this Madelung energy ordering to fill many-electrons atom.

Finally, from the points discussed above, we can now observe deeper a periodic table (see Figure 3.10). Note the lines represent the principal quantum number n. For $n = 1$, there is only one possible value of l, i.e., $l = 0$, and then $m = 0$. Thus, this shell has only one orbital and then two electrons (each orbital is able to receive two electrons: one with spin up and other with spin down). From this idea we see two elements: Hydrogen and Helium. Following the Madelung energy ordering, the next energy level is 2s, i.e., $n = 2$ and $l = 0$. Analogously to the first line, this orbital receive two electrons (Lithium and Beryllium). The next level is 2p and thus $n = 2$ and $l = 1$. This case has $2l + 1 = 3$ different orbitals, with, consequently, six electrons and then six elements: Boron, Carbon, Nitrogen, Oxygen, Fluorine, and Neon. This idea goes through the whole periodic table and the shells are then *aufbau*.

4 Thermodynamics

This chapter is devoted to remember some important concepts of thermodynamics; those needed to the aim of this book. Before going deeper, let us remember some definitions:

Thermodynamic system is any macroscopic system.

Thermodynamic parameter is any measurable macroscopic quantity.

Thermodynamic state is a set of thermodynamic parameters used to characterize a system.

Thermodynamic equilibrium occurs when the thermodynamic parameters do not change in value as a function of time.

Equation of state is a functional that provides a relationship between the thermodynamic parameters. Generally, we are looking for equation of states, to then know the evolution of a thermodynamic quantity as a function of other.

4.1 Thermodynamic laws

The thermodynamic laws are based on some concepts that can be expressed in different ways. Here, the aim of this description is not to provide a closed definition of those laws, but to introduce/review those.

Zero law states that there exists only one temperature for systems in thermal equilibrium.

First law ensures energy conservation. It can be expressed as follows: there is a thermodynamic potential named internal energy, and its change dU is equal to the amount of heat dQ supplied to the system plus the work dW done on the system:

$$dU = dW + dQ. \tag{4.1}$$

Second law forbids a perpetual machine. As an example, heat always flows spontaneously from the hotter sink to the colder one; and never the contrary. However, this second law is quite rich in concept and, as mentioned above, can be expressed from several points of view. In this sense, an additional information is important to remember. From Clausius theorem it is possible to define the (differential) entropy S:

$$dS = \frac{dQ}{T}, \tag{4.2}$$

Fundamentals of Magnetism. http://dx.doi.org/10.1016/B978-0-12-405545-2.00004-7

i.e., for an isothermal process, the entropy change dS is equal to the heat exchange dQ divided by the temperature of the system.

Third law states that the entropy tends to zero as the temperature approaches zero.

After this simple presentation of the thermodynamic laws, we can go further and consider the (magnetic) work as:

$$W = BM + \mu N, \tag{4.3}$$

where B is the magnetic field, M magnetization, μ chemical potential, and finally, N the number of particles in the system. Thus, the internal energy (from Eqs. (4.1) and (4.2)) resumes as:

$$U = TS + BM + \mu N. \tag{4.4}$$

4.2 Entropy

Let us consider the entropy S as a function of internal energy U, magnetization M, and number of particles N. Thus,

$$S = S(U, M, N). \tag{4.5}$$

From the first law of thermodynamics, we know that

$$T\, dS = dU - dW \tag{4.6}$$

and thus (considering Eq. (4.3)) we obtain one differential entropy:

$$T\, dS = dU - B\, dM - \mu\, dN. \tag{4.7}$$

However, we can obtain other differential entropy (from Eq. (4.5)):

$$dS = \frac{\partial S}{\partial U}\bigg|_{M,N} dU + \frac{\partial S}{\partial M}\bigg|_{U,N} dM + \frac{\partial S}{\partial N}\bigg|_{U,M} dN. \tag{4.8}$$

Comparing Eqs. (4.7) and (4.8) we obtain

$$1 = T\frac{\partial S}{\partial U}\bigg|_{M,N}, \quad -B = T\frac{\partial S}{\partial M}\bigg|_{U,N}, \quad -\mu = T\frac{\partial S}{\partial N}\bigg|_{U,M}. \tag{4.9}$$

Note that it is possible to define, from the equations above, some quantities: temperature T, chemical potential μ, and magnetic induction B.

4.3   Thermodynamic potentials

One of the main objectives of thermodynamic problems is to obtain the equation of state of the system under consideration. It is important because we can know the evolution of one thermodynamic parameter as a function of other. In this sense, we need to define some thermodynamic potentials, that, on its turn, help to obtain the equation of state of the system.

As will be clear, all of them are defined as a function of the internal energy U, and the choice of each potential is made by convenience, in accordance with the system studied.

4.3.1   Internal energy

Internal energy is a thermodynamic potential and let us consider that it depends on the entropy S, magnetization M, and number of particles N, analogously to the entropy. Thus:

$$U = U(S, M, N). \tag{4.10}$$

From Eq. (4.4) we can then write:

$$dU = T\,dS + B\,dM + \mu\,dN. \tag{4.11}$$

Analogously to what was done to the entropy (and will be done for the other thermodynamic potentials, below), we can also write other differential internal energy (from Eq. (4.10)):

$$dU = \left.\frac{\partial U}{\partial S}\right|_{M,N} dS + \left.\frac{\partial U}{\partial M}\right|_{S,N} dM + \left.\frac{\partial U}{\partial N}\right|_{S,M} dN \tag{4.12}$$

and then, comparing Eqs. (4.11) and (4.12), we can obtain the temperature T, magnetic induction B, and chemical potential μ, from the internal energy:

$$T = \left.\frac{\partial U}{\partial S}\right|_{M,N}, \quad B = \left.\frac{\partial U}{\partial M}\right|_{S,N}, \quad \mu = \left.\frac{\partial U}{\partial N}\right|_{S,M}. \tag{4.13}$$

4.3.2   Helmholtz free energy

Helmholtz free energy is defined as

$$F = U - TS \tag{4.14}$$

and then, the differential one is

$$dF = dU - T\,dS - S\,dT \tag{4.15}$$

and, consequently, considering Eq. (4.11), we can rewrite the above differential expression as

$$dF = B\,dM + \mu\,dN - S\,dT. \tag{4.16}$$

It is straightforward to see that the Helmholtz free energy depends on the temperature T, magnetization M, and number of particles N:

$$F = F(T, M, N) \tag{4.17}$$

and then, following the procedure done for the entropy and internal energy, we can write:

$$dF = \left.\frac{\partial F}{\partial T}\right|_{M,N} dT + \left.\frac{\partial F}{\partial M}\right|_{T,N} dM + \left.\frac{\partial F}{\partial N}\right|_{T,M} dN. \tag{4.18}$$

Thus, we can obtain, from the Helmholtz free energy, the entropy S, chemical potential μ, and magnetic induction B:

$$-S = \left.\frac{\partial F}{\partial T}\right|_{M,N}, \qquad \mu = \left.\frac{\partial F}{\partial N}\right|_{T,M}, \qquad B = \left.\frac{\partial F}{\partial M}\right|_{T,N}. \tag{4.19}$$

4.3.3   Enthalpy

Enthalpy is defined as

$$H = U - BM \tag{4.20}$$

and then, the differential form is:

$$dH = dU - B\,dM - M\,dB. \tag{4.21}$$

Again, considering Eq. (4.11), the differential form is:

$$dH = T\,dS + \mu\,dN - M\,dB. \tag{4.22}$$

Thus, the enthalpy depends on the entropy S, magnetic induction B, and number of particles N:

$$H = H(S, B, N). \tag{4.23}$$

The differential enthalpy obtained from the above equation is

$$dH = \left.\frac{\partial H}{\partial S}\right|_{B,N} dS + \left.\frac{\partial H}{\partial B}\right|_{S,N} dB + \left.\frac{\partial H}{\partial N}\right|_{S,B} dN \tag{4.24}$$

and then, this thermodynamic potential defines the temperature T, chemical potential μ, and magnetization M:

$$T = \left.\frac{\partial H}{\partial S}\right|_{B,N}, \qquad \mu = \left.\frac{\partial H}{\partial N}\right|_{S,B}, \qquad -M = \left.\frac{\partial H}{\partial B}\right|_{S,N}. \tag{4.25}$$

4.3.4 Gibbs free energy (free enthalpy)

This thermodynamic potential is defined as:

$$G = U - TS - BM \qquad (4.26)$$

and then:

$$dG = dU - S\,dT - T\,dS - B\,dM - M\,dB \qquad (4.27)$$

and from Eq. (4.11) we obtain

$$dG = \mu\,dN - S\,dT - M\,dB. \qquad (4.28)$$

From the above evaluation we see the thermodynamic dependence of the Gibbs free energy: temperature T, magnetic induction B, and number of particles N:

$$G = G(T, B, N). \qquad (4.29)$$

The differential form of the above equation is

$$dG = \left.\frac{\partial G}{\partial T}\right|_{B,N} dT + \left.\frac{\partial G}{\partial B}\right|_{T,N} dB + \left.\frac{\partial G}{\partial N}\right|_{T,B} dN \qquad (4.30)$$

and then, from the Gibbs free energy, we obtain the entropy S, chemical potential μ, and magnetization M:

$$-S = \left.\frac{\partial G}{\partial T}\right|_{B,N}, \qquad \mu = \left.\frac{\partial G}{\partial N}\right|_{T,B}, \qquad -M = \left.\frac{\partial G}{\partial B}\right|_{T,N}. \qquad (4.31)$$

4.3.5 Grand potential

The last thermodynamic potential discussed here is the *Grand* potential, defined as

$$\Phi = U - TS - \mu N \qquad (4.32)$$

with a differential form as below:

$$d\Phi = dU - S\,dT - T\,dS - \mu\,dN - N\,d\mu \qquad (4.33)$$

that, on its turn, can be rewritten as (with the help of Eq. (4.11)):

$$d\Phi = B\,dM - S\,dT - N\,d\mu. \qquad (4.34)$$

From the above equation it is possible to see the thermodynamic dependence of the *Grand* potential:

$$\Phi = \Phi(T, M, \mu) \qquad (4.35)$$

and then we obtain the other differential form:

$$d\Phi = \left.\frac{\partial \Phi}{\partial T}\right|_{M,\mu} dT + \left.\frac{\partial \Phi}{\partial M}\right|_{T,\mu} dM + \left.\frac{\partial \Phi}{\partial \mu}\right|_{T,M} d\mu. \tag{4.36}$$

Finally, from this thermodynamic potential, we can obtain the entropy S, magnetic induction B, and number of particles N:

$$-S = \left.\frac{\partial \Phi}{\partial T}\right|_{M,\mu}, \qquad B = \left.\frac{\partial \Phi}{\partial M}\right|_{T,\mu}, \qquad -N = \left.\frac{\partial \Phi}{\partial \mu}\right|_{T,M}. \tag{4.37}$$

4.4 Maxwell relationships

The Maxwell relationships are quite useful, because these provide a simple relation between the thermodynamic quantities. Let us obtain one Maxwell relationship as an example. To this purpose, consider Eqs. (4.13)-left and center. Note one is derived with respect to S and other with respect to M. Thus, the procedure is to derive with respect to M that one with S-derivative and, analogously, derive with respect to S that one with M-derivative. Thus,

$$\left.\frac{\partial T}{\partial M}\right|_{S,N} = \left.\frac{\partial}{\partial M}\left(\left.\frac{\partial U}{\partial S}\right|_{M,N}\right)\right|_{S,N} \tag{4.38}$$

and

$$\left.\frac{\partial B}{\partial S}\right|_{M,N} = \left.\frac{\partial}{\partial S}\left(\left.\frac{\partial U}{\partial M}\right|_{S,N}\right)\right|_{M,N}. \tag{4.39}$$

Now, we need to remember an important property:

$$\frac{\partial^2}{\partial x \partial y} = \frac{\partial^2}{\partial y \partial x} \tag{4.40}$$

and from Eqs. (4.38) and (4.39) we obtain

$$\left.\frac{\partial T}{\partial M}\right|_{S,N} = \left.\frac{\partial B}{\partial S}\right|_{M,N}. \tag{4.41}$$

This is only one of the several Maxwell relations. Some of them are summarized in the next section.

4.5 Thermodynamic square

The thermodynamic square (valid only for a constant number of particles N) is an interesting way to summarize (i) the derivation of thermodynamic quantities from thermodynamic potentials and (ii) Maxwell relations.

It works as follows: on the sides and vertices of the square we find the thermodynamic potentials and quantities, respectively. The derivative of one thermodynamic potential with respect to the next neighbor quantity is equal to the quantity in the opposite side of the square. Note if the arrow follows the direction of the opposite side, the relation is positive; otherwise, it is negative. For instance, we can recover

$$\frac{\partial F}{\partial T} = -S. \tag{4.42}$$

The thermodynamic square also provides the Maxwell relations. The derivative of one thermodynamic quantity with respect to the next neighbor quantity (with constant next-next neighbor) is equal to the derivative of the other next neighbor quantity with respect to the next-next neighbor quantity (with constant next-next-next neighbor quantity). Complicated? It is not. Verify the following:

$$\frac{\partial M}{\partial T}\bigg|_B = \frac{\partial S}{\partial B}\bigg|_T. \tag{4.43}$$

4.6 Magnetic specific heat

Specific heat is an important quantity and must be here briefly introduced. It is defined as the quantity of heat needed to change the temperature of the system in a certain amount:

$$C_i = \frac{dQ}{dT}\bigg|_i, \tag{4.44}$$

where i is a constant parameter. From Eq. (4.7), we know that:

$$dQ = T\,dS = dU - B\,dM - \mu\,dN \tag{4.45}$$

and this expression is important to the evaluations below, considering N constant ($dN = 0$).

The specific heat under constant magnetization assumes $dM = 0$ and then $dQ = T\,dS = dU$. Thus,

$$C_M = \frac{dQ}{dT}\bigg|_M = T\frac{\partial S}{\partial T}\bigg|_M = \frac{\partial U}{\partial T}\bigg|_M \tag{4.46}$$

and then C_M depends on the internal energy U.

The specific heat under constant magnetic field requires $dB = 0$ and then $dQ = T\,dS = dU - B\,dM = dH$. Thus:

$$C_B = \frac{dQ}{dT}\bigg|_B = T\frac{\partial S}{\partial T}\bigg|_B = \frac{\partial H}{\partial T}\bigg|_B \tag{4.47}$$

and then C_B depends on the enthalpy H.

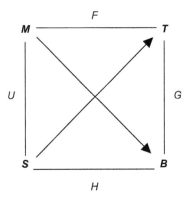

Figure 4.1 Thermodynamic square for constant number of particles N. From this square it is possible to obtain the thermodynamic quantities with respect to the thermodynamic potentials and the Maxwell relations (see text for further details).

There are relationships between C_M and C_B, for constant N, as presented below. A detailed proof of these is left as an exercise to the reader (see Figure 4.1):

$$C_B = C_M + T \left[\frac{\partial M}{\partial T} \bigg|_B \right]^2 \left[\frac{\partial M}{\partial B} \bigg|_T \right]^{-1}, \tag{4.48}$$

$$\frac{C_B}{C_M} = \frac{\frac{\partial M}{\partial B} \big|_T}{\frac{\partial M}{\partial B} \big|_S}. \tag{4.49}$$

5 Statistical Mechanics

The connection between the micro- and macroscopic worlds lies on the entropy functional; more precisely on the Shannon entropy:

$$s = -\sum_i p_i \ln p_i \,, \tag{5.1}$$

where p_i is the equilibrium probability to find the system in a certain level i. Physically speaking, the above entropy assumes the correct dimension considering $S = k_B s$.

The question that arises is quite intuitive: what are those probabilities? To obtain those, we need to consider the maximum entropy principle, which states that the equilibrium distribution is found when the entropy is maximized under the correct constraint. It is then obtained considering:

$$\frac{ds}{dp_i} = 0 \tag{5.2}$$

and the Lagrange multiplier technique (see next sections for details).

Then, after obtaining the equilibrium distribution p_i, it is possible to determine the mean value of thermodynamic quantities. To this purpose, we consider:

$$\mathcal{O} = \sum_i p_i \mathcal{O}_i. \tag{5.3}$$

From now on, let us discuss three different ensembles; three different constraints to the entropy and then determine the corresponding equilibrium distribution. The first ensemble discussed is named *micro canonical* and represents an isolated system, with constant number of particles and energy. The second ensemble considered is called *canonical* and, for this case, the system is in thermal contact with a thermal reservoir; can exchange energy, but the temperature remains constant. Note, for this case, we need to add an extra constraint to the system, that rules energy fluctuations. Finally, the third ensemble considered is named *Grand canonical* and, in addition to exchange energy with a thermal reservoir, it can exchange particles with a particle reservoir. Analogously to before, we must add an extra constraint to describe this case and this one rules particles fluctuations. Table 5.1 summarizes these ideas.

Fundamentals of Magnetism. http://dx.doi.org/10.1016/B978-0-12-405545-2.00005-9

Table 5.1 Ensembles considered in this book and their main characteristics.

Ensemble	Characteristic	Fixed Variables
Micro canonical	Isolated	U, N
Canonical	Heat exchange	N
Grand canonical	Heat and particles exchange	–

5.1 Micro canonical ensemble

As qualitatively discussed above, this ensemble is isolated and has a fixed energy and number of particles. Thus, there is only one constraint to the entropy:

$$\sum_i p_i = 1. \tag{5.4}$$

This one ensures the normalization of the distribution probability p_i and, to consider the Lagrange multiplier technique, the above equation must be rewritten as

$$1 - \sum_i p_i = 0. \tag{5.5}$$

This constraint is then ready to be added to the entropy, multiplied by the Lagrange multiplier λ:

$$s = -\sum_i p_i \ln p_i + \lambda \left(1 - \sum_i p_i \right). \tag{5.6}$$

Considering the maximum entropy principle, we must follow:

$$\frac{ds}{dp_i} = 0 \tag{5.7}$$

to determine the equilibrium distribution p_i. After a simple calculation we obtain

$$p_i = e^{-(1+\lambda)} = C, \tag{5.8}$$

where C is a constant. Thus, the above result back into the entropy we find

$$p_i = \frac{1}{\Omega}, \tag{5.9}$$

where $\Omega = \sum_i$ is the number of accessible states of the system. Note this result means that the available levels are equiprobables, each one with probability $1/\Omega$ to be found. A good example is a dice game, where $\Omega = 6$, $i = 1, \ldots, 6$, and $p_i = 1/6$.

We can go further and introduce the above result back again into the entropy. Then, we find:

$$S = k_B \ln \Omega \tag{5.10}$$

the famous Boltzmann entropy.

5.2 Canonical ensemble

This ensemble, by definition, is in thermal contact with a thermal reservoir. It then exchanges heat with this reservoir, but its temperature remains constant. Thus, we need to add a new constraint to the entropy, in such a way that it permits the system to change the energy of each state. The new constraint is the mean energy:

$$U = \sum_i p_i \varepsilon_i \tag{5.11}$$

and, analogously to before, can be added as zero to the entropy:

$$s = -\sum_i p_i \ln p_i + \lambda \left(1 - \sum_i p_i \right) + \beta \left(U - \sum_i p_i \varepsilon_i \right). \tag{5.12}$$

Considering the maximum entropy principle, the equilibrium distribution is

$$p_i = C' e^{-\beta \varepsilon_i}, \tag{5.13}$$

where C' is a constant. As before, the above expression can be placed into the normalization condition to p_i (Eq. (5.4)). This procedure allows us to determine C' and then, the final expression to the probability distribution of the levels:

$$p_i = \frac{e^{-\beta \varepsilon_i}}{Z}, \tag{5.14}$$

where

$$Z = \sum_i e^{-\beta \varepsilon_i} \tag{5.15}$$

is the partition function. Note, as required on the definition of this canonical ensemble, that the probability to find the system in a certain level i depends on the energy of the level.

From now on, we will write the thermodynamic quantities (and potential), in microscopic terms, i.e., depending on the energy levels, via partition function Z.

5.2.1 Entropy

Analogously to before, Eq. (5.14) can be placed back into the entropy functional and then we find:

$$S = k_B (\beta U + \ln Z) \tag{5.16}$$

that means the equilibrium entropy.

Now, an important question arises: what is the physical meaning of the Lagrange multiplier β? Let us then go back to Eq. (4.9)-left:

$$\frac{\partial S}{\partial U} = \frac{1}{T}.$$

(5.17)

Thus, the derivative of Eq. (5.16) with respect to the mean energy leads to:

$$\beta = \frac{1}{k_B T}.$$

(5.18)

Note the system has a constant temperature T.

5.2.2 Helmholtz free energy

This potential is of great importance to magnetism, since from this, we can obtain the magnetization. From Eq. (4.14), we know that

$$F = U - TS$$

(5.19)

and then, after placing Eq. (5.16) into the above equation, we obtain

$$F = -k_B T \ln Z \; .$$

(5.20)

Let us focus our attention back to the entropy. Considering Eq. (4.19)-left, i.e.,

$$S = -\frac{\partial F}{\partial T},$$

(5.21)

we are able to write a new expression to the entropy, depending on the partition function:

$$S = k_B \frac{\partial}{\partial T} (T \ln Z) \; .$$

(5.22)

5.2.3 Mean energy

The mean energy is given in accordance with Eq. (5.11):

$$U = \frac{\sum_i \varepsilon_i e^{-\beta \varepsilon_i}}{\sum_i e^{-\beta \varepsilon_i}}.$$

(5.23)

It is easy to verify that

$$U = -\frac{1}{Z} \frac{\partial Z}{\partial \beta}$$

(5.24)

and then, in a closed and friendly form:

$$U = k_B T^2 \frac{\partial}{\partial T} \ln Z \; .$$

(5.25)

5.2.4 Mean magnetic moment

This quantity is quite important, since it is the aim of this book. Let us start then from the (Zeeman) energy of the level i. Of course, there are other important energies to the magnetism, however, we are considering here only that depending on the magnetic field of induction $\vec{B}$. Thus:

$$\varepsilon_i = -\vec{\mu}_i \cdot \vec{B} = -\sum_{u=x,y,z} (\mu_i)_u B_u. \tag{5.26}$$

Note, and this information will be quite important further, the magnetic moment of the level i can also be defined from:

$$(\mu_i)_u = -\frac{\partial \varepsilon_i}{\partial B_u}. \tag{5.27}$$

Thus, the mean magnetic moment $\vec{\mu}$ can be obtained as

$$\begin{aligned}
\vec{\mu} &= \sum_i p_i \vec{\mu}_i \\
&= \sum_i p_i (\mu_i)_x \hat{x} + \sum_i p_i (\mu_i)_y \hat{y} + \sum_i p_i (\mu_i)_z \hat{z} \\
&= \mu_x \hat{x} + \mu_y \hat{y} + \mu_z \hat{z}.
\end{aligned} \tag{5.28}$$

We must then determine μ_u:

$$\begin{aligned}
\mu_u &= \sum_i p_i (\mu_i)_u \\
&= \frac{\sum_i (\mu_i)_u \exp\left[\sum_u (\mu_i)_u B_u \beta\right]}{\sum_i \exp\left[\sum_u (\mu_i)_u B_u \beta\right]}.
\end{aligned} \tag{5.29}$$

Analogously to the mean energy above discussed, it is possible to write the above equation as:

$$\mu_u = \frac{1}{\beta Z} \frac{\partial Z}{\partial B_u} = -\frac{\partial F}{\partial B_u}, \tag{5.30}$$

where $u = x, y, z$. Finally, the above equation defines the mean magnetic moment μ_u as a function of the partition function (and the Helmholtz free energy).

5.3 Indistinguishable and distinguishable particles

Now, an important question arises: how do those thermodynamic quantities scale for N particles? Depends on the character of the particles. These can be either distinguishable or indistinguishable. For instance, particles in a lattice are distinguishable; while in a quantum gas, those particles are indistinguishable.

(a) Distinguishable particles.   (b) Indistinguishable particles.

Figure 5.1 Possible distributions of two (a) distinguishable and (b) indistinguishable particles in a two-levels system.

To determine this scale for N-particle system, let us consider a two-levels system, with two distinguishable particles, as presented in Figure 5.1a.

Note there are four possible ways to fill the levels. Thus, the total canonical partition function is

$$Z(2) = e^0 + 2e^{-\varepsilon\beta} + e^{-2\varepsilon\beta} \tag{5.31}$$

that can be rewritten as

$$Z(2) = \left(1 + e^{-\varepsilon\beta}\right)^2 = \left[Z(1)\right]^2. \tag{5.32}$$

As can be seen, the partition function of the 2-particle system depends on the partition function of the single particle system. We can then generalize this result for *distinguishable* particles as:

$$Z(N) = \left[Z(1)\right]^N. \tag{5.33}$$

The scenario is a bit different for indistinguishable particles, as drawn in Figure 5.1b. For this case, there are only three possible combinations to distribute the two indistinguishable particles into those two levels available. The partition function is then:

$$Z(2) = e^0 + e^{-\varepsilon\beta} + e^{-2\varepsilon\beta} \neq \left[Z(1)\right]^2. \tag{5.34}$$

Note the scale to the N-particles system can no longer be generalized as before; where $\left[Z(1)\right]^N$ overcounts ($N!$ times), the distributions in which all N particles occupy different levels. Considering the low density limit, i.e., there are much more levels available than particles; then, we can assume that these distributions (one particle for each level) have quite low probability. Thus, it is possible to write, for *indistinguishable* particles:

$$Z(N) = \frac{\left[Z(1)\right]^N}{N!}. \tag{5.35}$$

5.4 Thermodynamic quantities of *N* distinguishable particles

The aim of the present section is to obtain the thermodynamic quantities, in the canonical ensemble, for N distinguishable particles. Thus, Eq. (5.33) rules the development here presented. Starting from the entropy, we must recall Eq. (5.22) and then:

$$S(N) = Nk_B \frac{\partial}{\partial T}\{T \ln[Z(1)]\} = NS(1). \tag{5.36}$$

Analogously, we can obtain the Helmholtz free energy (from Eq. (5.20)) and the mean energy (from Eq. (5.25)), respectively:

$$F(N) = -Nk_B T \ln[Z(1)] = NF(1) \tag{5.37}$$

and

$$U(N) = Nk_B T^2 \frac{\partial}{\partial T} \ln[Z(1)] = NU(1). \tag{5.38}$$

In what concerns the mean magnetic moment, it scales as (from Eq. (5.30))

$$\mu_u(N) = -\frac{\partial F(N)}{\partial B_u} = -N\frac{\partial F(1)}{\partial B_u} = N\mu_u(1) \tag{5.39}$$

and then we can write the magnetization

$$M_u = \mu_u(N) = N\mu_u(1). \tag{5.40}$$

Finally, we can consider the magnetic susceptibility, defined as

$$\chi = \lim_{H \to 0} \frac{\partial M}{\partial H} = \lim_{B \to 0} \mu_0 \frac{\partial M}{\partial B}, \tag{5.41}$$

where $B = \mu_0 H$. Note both, magnetization and magnetic induction are vectors and then the magnetic susceptibility must be a *rank*-2 tensor. Thus, the most correct form to write the magnetic susceptibility is:

$$\chi_{vu} = \lim_{B_v \to 0} \mu_0 \frac{\partial M_u}{\partial B_v}, \tag{5.42}$$

where, as an example, we have

$$\chi_{zy} = \lim_{B_z \to 0} \mu_0 \frac{\partial M_y}{\partial B_z}. \tag{5.43}$$

Finally, the magnetic susceptibility can be written as

$$\chi = \begin{pmatrix} \chi_{xx} & \chi_{xy} & \chi_{xz} \\ \chi_{yx} & \chi_{yy} & \chi_{yz} \\ \chi_{zx} & \chi_{zy} & \chi_{zz} \end{pmatrix}. \tag{5.44}$$

5.5 *Grand* canonical ensemble

Analogously to before, now we need to introduce a new constraint, related to the mean number of particles N. This constraint is

$$N = \sum_i p_i n_i, \tag{5.45}$$

where n_i is the number of particles in the ith state. Following the Lagrange multiplier technique (as before), this new information must be added to the entropy functional:

$$s = -\sum_i p_i \ln p_i + \lambda \left(1 - \sum_i p_i\right) + \beta \left(U - \sum_i p_i \varepsilon_i\right) + \gamma \left(N - \sum_i p_i n_i\right) \tag{5.46}$$

and after the maximum entropy principle we obtain

$$p_i = \mathcal{C}'' e^{-\beta \varepsilon_i} e^{-\gamma n_i}. \tag{5.47}$$

We need to determine $\mathcal{C}''$, a simple task done after place the above equation back into the normalization condition to the distribution probabilities (Eq. (5.4)). After that, the above equation reads as

$$p_i = \frac{e^{-\beta \varepsilon_i} e^{-\gamma n_i}}{\mathcal{Z}}, \tag{5.48}$$

where

$$\mathcal{Z} = \sum_i e^{-\beta \varepsilon_i} e^{-\gamma n_i}. \tag{5.49}$$

Now, we have the same problem as before to solve: determine the physical meaning of the Lagrange multipliers β and γ. To go further, let us go back the entropy.

5.5.1 Entropy

If we place back the equilibrium distribution obtained above (Eq. (5.48)), into the Shannon entropy (Eq. (5.1)) we obtain (analogously to the canonical ensemble):

$$S = k_B(\beta U + \gamma N + \ln \mathcal{Z}). \tag{5.50}$$

We must now recover Eqs. (4.9)-left and right:

$$\frac{\partial S}{\partial U} = \frac{1}{T} \quad \text{and} \quad -\mu = T\frac{\partial S}{\partial N} \tag{5.51}$$

to then obtain

$$\beta = \frac{1}{k_B T} \tag{5.52}$$

and

$$\gamma = -\mu\beta. \tag{5.53}$$

Note β has the same meaning as before (canonical case), and rules the energy exchange with the thermal reservoir; while the Lagrange parameter γ is related to the chemical potential μ and, consequently, rules the particles exchange with the particle reservoir.

5.5.2 Grand potential

For the canonical case, we considered the Helmholtz free energy F, however, now the best thermodynamic potential to consider is the *Grand* potential, since it includes the mean number of particles N (from Eq. (4.32)):

$$\Phi = U - TS - \mu N. \tag{5.54}$$

Considering the entropy as obtained above (Eq. (5.50)) we obtain

$$\Phi = -k_{\rm B} T \ln \mathcal{Z} \tag{5.55}$$

analogously to the Helmholtz free energy. For the sake of simplicity, we will write the above equation as

$$\Phi = -\frac{1}{\beta} q, \tag{5.56}$$

where

$$q = \ln \mathcal{Z}. \tag{5.57}$$

We can return back to the entropy and rewrite that, considering the following thermodynamic equation (see Eq. (4.37)):

$$S = -\frac{\partial \Phi}{\partial T}. \tag{5.58}$$

Then:

$$S = k_{\rm B} \frac{\partial}{\partial T} (T \ln \mathcal{Z}). \tag{5.59}$$

5.5.3 Magnetization

Analogously to before, one component of the vector magnetization is

$$M_u = \frac{1}{\mathcal{Z}} \sum_i (\mu_i)_u \exp\left\{\left[\sum_u (\mu_i)_u B_u + \mu n_i\right]\beta\right\} \tag{5.60}$$

and then

$$M_u = \frac{1}{\beta} \frac{\partial}{\partial B_u} \ln \mathcal{Z}. \tag{5.61}$$

In terms of the *Grand* potential, it is

$$M_u = -\frac{\partial \Phi}{\partial B_u}. \tag{5.62}$$

Note $(\mu_i)_u$ is related to the magnetic moment along u direction, while μ is the chemical potential.

5.5.4   Mean number of particles

Finally, the mean number of particles into the ensemble is

$$N = \sum_i n_i p_i \tag{5.63}$$

that can be written as

$$N = \frac{1}{\mathcal{Z}} \sum_i n_i e^{-\varepsilon_i \beta} e^{\mu n_i \beta}. \tag{5.64}$$

In terms of the *Grand* potential it resumes as

$$N = -\frac{\partial \Phi}{\partial \mu}. \tag{5.65}$$

Note, as expected, this result is the same as that obtained from thermodynamic arguments (Eq. (4.37)).

5.6   *Grand* partition function: further developments

Before we go further, let us consider the occupation number representation $\{n_k\}$. For bosons, there is no limit of occupation in the same quantum level, and then: $n_k = 1, 2, 3, \ldots, N$; on the other hand, for fermions, due to the Pauli exclusion principle, there is a limit to this occupation. Thus, for this last case: $n_k = 0, 1$.

We can also write the total energy E and total number of particles $\mathcal{N}$ in terms of the occupation number n_k:

$$E = \sum_k n_k \varepsilon_k \tag{5.66}$$

and

$$\mathcal{N} = \sum_k n_k. \tag{5.67}$$

From these considerations, it is not difficult to write (based on Eq. (5.49)), the N-particles *Grand* partition function in terms of the occupation number representation:

$$\mathcal{Z} = \sum_{\{n_k\}} g\{n_k\} \exp\left[-\beta \sum_{k=1}^{+\infty} n_k(\epsilon_k - \mu)\right], \qquad (5.68)$$

where $g\{n_k\}$ is the statistical weight of the occupation number; and assumes $g\{n_k\} = \frac{1}{n_1!}\frac{1}{n_2!}\cdots$ for the Maxwell–Boltzmann (MB) statistics and $g\{n_k\} = 1$ for both, Fermi–Dirac (FD) and Bose–Einstein (BE) statistics.

From now on, let us obtain a friendly expression to each of the statistics. Considering first the Fermi–Dirac one, it is possible to follow the steps given below:

$$
\begin{aligned}
\mathcal{Z}^{\mathrm{FD}} &= \sum_{n_1,n_2,\ldots=0}^{1} \exp[-\beta n_1(\epsilon_1 - \mu)]\exp[-\beta n_2(\epsilon_2 - \mu)]\cdots \\
&= \sum_{n_1,n_2,\ldots=0}^{1} \{\exp[-\beta(\epsilon_1 - \mu)]\}^{n_1}\{\exp[-\beta(\epsilon_2 - \mu)]\}^{n_2}\cdots \\
&= \prod_{k=1}^{+\infty}\sum_{n_k=0}^{1} \{\exp[-\beta(\epsilon_k - \mu)]\}^{n_k} \\
&= \prod_{k=1}^{+\infty}(1 + z\,e^{-\beta\epsilon_k}). \qquad (5.69)
\end{aligned}
$$

Note above, n_k can only assume two possible values: 0 and 1 (condition unique to the FD statistics); and $z = e^{\mu\beta}$ is the fugacity.

The same idea as above can be used to the Bose–Einstein statistics:

$$
\begin{aligned}
\mathcal{Z}^{\mathrm{BE}} &= \sum_{n_1,n_2,\ldots=0}^{+\infty} \{\exp[-\beta(\epsilon_1 - \mu)]\}^{n_1}\{\exp[-\beta(\epsilon_2 - \mu)]\}^{n_2}\cdots \\
&= \prod_{k=1}^{+\infty}\sum_{n_k=0}^{+\infty} \{\exp[-\beta(\epsilon_k - \mu)]\}^{n_k} \\
&= \prod_{k=1}^{+\infty}\frac{1}{1 - z\,e^{-\beta\epsilon_k}}. \qquad (5.70)
\end{aligned}
$$

Note that above n_k can assume different values; and now, due to the quantum properties of a bosons gas, n_k runs from zero up to infinity. In addition, the sum of the second line has a well-known solution.[1]

[1] For $x < 1$: $\sum_{n=0}^{+\infty} x^n = \frac{1}{1-x}$.

Finally, for the Maxwell–Boltzmann case, the same can be done:

$$
\begin{aligned}
\mathcal{Z}^{MB} &= \sum_{n_1,n_2,\dots=0}^{+\infty} \frac{1}{n_1!}\{\exp[-\beta(\epsilon_1-\mu)]\}^{n_1} \frac{1}{n_2!}\{\exp[-\beta(\epsilon_2-\mu)]\}^{n_2}\cdots \\
&= \prod_{k=1}^{+\infty}\sum_{n_k=0}^{+\infty} \frac{1}{n_k!}\{\exp[-\beta(\epsilon_k-\mu)]\}^{n_k} \\
&= \prod_{k=1}^{+\infty} \exp[z\,e^{-\beta\epsilon_k}].
\end{aligned}
\tag{5.71}
$$

Analogously to the previous case, the sum of the second line above has a well-defined solution.[2]

It is easy to obtain $q = \ln\mathcal{Z}$ for the three studied cases: FD, BE, and MB. Thus:

$$
\begin{aligned}
q^{FD} &= \ln\left\{\prod_{k=1}^{+\infty}\left(1+z\,e^{-\beta\epsilon_k}\right)\right\} \\
&= \sum_{k=1}^{+\infty}\ln\left(1+z\,e^{-\beta\epsilon_k}\right),
\end{aligned}
\tag{5.72}
$$

$$
\begin{aligned}
q^{BE} &= \ln\left\{\prod_{k=1}^{+\infty}\frac{1}{1-z\,e^{-\beta\epsilon_k}}\right\} \\
&= -\sum_{k=1}^{+\infty}\ln\left(1-z\,e^{-\beta\epsilon_k}\right),
\end{aligned}
\tag{5.73}
$$

$$
\begin{aligned}
q^{MB} &= \ln\left\{\prod_{k=1}^{+\infty}\exp[z\,e^{-\beta\epsilon_k}]\right\} \\
&= \sum_{k=1}^{+\infty} z\,e^{-\beta\epsilon_k}.
\end{aligned}
\tag{5.74}
$$

And, for the sake of completeness, these results can be summarized as

$$
q = \frac{1}{a}\sum_{k=1}^{+\infty}\ln\left(1+az\,e^{-\beta\epsilon_k}\right),
\tag{5.75}
$$

where $a = -1, +1$ for the BE and FD cases, respectively. The MB case is recovered considering the limit $a \to 0$.

---

[2]For $x \to 0$: $\sum_{n=0}^{+\infty}\frac{1}{n!}x^n = e^x$.

6 Fermions Gas

The results derived in this chapter are of great importance to develop the (dia-, para-, and ferro-) magnetic behavior of an electron gas. However, why do we need to know the magnetic behavior of an electron gas? Because some materials, like those with metallic character, can be described from this approximation.

6.1 Wave function, eigenvalues, and density of states

Let us therefore consider an electron gas, closed in a box with volume $V = L^3$. To describe this system we need to recall the time independent Schrödinger equation:

$$\mathcal{H}\Psi = E\Psi.$$

Considering only the kinetic energy of electrons, the Hamiltonian can be written as

$$\mathcal{H} = \frac{\vec{p}^2}{2m}, \tag{6.1}$$

where

$$\vec{p} = -i\hbar\vec{\nabla} \tag{6.2}$$

is the linear momentum operator. The Schrödinger equation can then be rewritten as:

$$-\frac{\hbar^2}{2m}\left(\frac{d^2}{dx^2} + \frac{d^2}{dy^2} + \frac{d^2}{dz^2}\right)\Psi(x, y, z) = E\Psi(x, y, z). \tag{6.3}$$

We found thus a differential equation and the solution for the wave function is a plane wave:

$$\Psi(x, y, z) = e^{ik_x x}\, e^{ik_y y}\, e^{ik_z z} = e^{i\vec{k}\cdot\vec{r}}, \tag{6.4}$$

where $\vec{k}$ defines the wave vector; and the energy eigenvalues are

$$E = \frac{\hbar^2}{2m}\left(k_x^2 + k_y^2 + k_z^2\right) = \frac{\hbar^2\vec{k}^2}{2m}. \tag{6.5}$$

Fundamentals of Magnetism. http://dx.doi.org/10.1016/B978-0-12-405545-2.00006-0

Well, those electrons are confined in a closed box and therefore the wave function $\Psi(x, y, z)$ must satisfy some boundary conditions. Let us consider the Born-von Karman one:

$$\Psi(x + L, y, z) = \Psi(x, y, z) \tag{6.6}$$

(and, of course, the same for the other two coordinates x and y). Thus, we obtain

$$e^{ik_x L} = e^{ik_y L} = e^{ik_z L} = 1 \tag{6.7}$$

and, consequently,

$$k_x = \frac{2\pi}{L} n_x, \quad k_y = \frac{2\pi}{L} n_y, \quad k_z = \frac{2\pi}{L} n_z, \tag{6.8}$$

where $n_u = 0, 1, 2, \ldots$ assumes only integer values ($u = x, y, z$).

From the above result it is also possible to rewrite the energy, since this quantity also depends on n_u:

$$E = \frac{\hbar^2}{2m} \frac{(2\pi)^2}{L^2} \left(n_x^2 + n_y^2 + n_z^2 \right). \tag{6.9}$$

Note that this simple system has a remarkable level of degeneracy, i.e., there are several sets of $\{n_x, n_y, n_z\}$ for the same energy E. For instance, the states, $\{\pm 1, 0, 0\}$, $\{0, \pm 1, 0\}$, and $\{0, 0, \pm 1\}$, all of them have the same energy

$$E = \frac{\hbar^2}{2m} \frac{(2\pi)^2}{L^2}. \tag{6.10}$$

There is another factor that increases the degeneracy of those states: the electron spin. There are two possible configurations for the spins in the same sates $\{n_x, n_y, n_z\}$, and these are $s = \pm 1/2$. Thus, for the example described above, it is possible to organize 12 spins (instead of 6) for the same energy of Eq. (6.10).

Since this system has degeneracy, let us obtain its density of states $g(E)$, i.e., the number N_k of possible states per energy E. This derivation will be done from purely geometric arguments and, for this effect, see Figure 6.1, that represents the k-space, i.e., the reciprocal space. The elemental volume V_e in this space is (from Eq. (6.8))

$$V_e = \left(\frac{2\pi}{L} \right)^3, \tag{6.11}$$

while the volume that a certain state $\vec{k}$ fills is

$$V_k = \frac{4}{3} \pi k^3. \tag{6.12}$$

Thus, the number of states N_k within a sphere with radius $|\vec{k}|$ is

$$N_k = \frac{V_k}{V_e} = \frac{4}{3} \pi \left(\frac{kL}{2\pi} \right)^3. \tag{6.13}$$

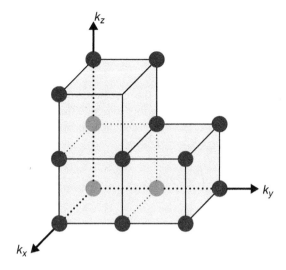

Figure 6.1 Reciprocal space representing some possible (and quantized) states; those described by Eq. (6.8).

Well, the density of states is the number of states per energy and must be written as

$$g(E) = \frac{dN_k}{dE} = \frac{dN_k}{dk}\frac{dk}{dE}. \tag{6.14}$$

Considering $E = \hbar^2 \vec{k}^2/2m$, we obtain

$$g(E) = \frac{\bar{g}V}{4\pi^2}\left(\frac{2m}{\hbar^2}\right)^{3/2} E^{1/2}, \tag{6.15}$$

where the factor $\bar{g} = 2s + 1$ was included "by hand" and is concerned with the multiplicity due to the particle spin (for the present case, i.e., an electron, $\bar{g} = 2$).

6.2 *Grand* canonical potential and thermodynamic quantities

Let us recall Eq. (5.57), where we defined q as the logarithm of the *grand* partition function. For the fermions case (Eq. (5.72)):

$$q = \ln \mathcal{Z} = \sum_{k=1}^{+\infty} \ln\left(1 + z\,e^{-\beta\epsilon_k}\right). \tag{6.16}$$

For large values of the wave vector $|\vec{k}|$ (i.e., large volumes in the reciprocal space), the above sum can be changed by an integral. Mathematically speaking, this transformation can be done using the concepts of Riemann's Integrals, and therefore:

$$\sum_k \rightarrow \left(\frac{L}{2\pi}\right)^3 \int d^3k. \tag{6.17}$$

Note that the prefactor of the integral is the inverse of the elemental volume V_e of the reciprocal space. Now, let us change this integral (that is actually in Cartesian coordinates), to spherical coordinates[1] and then change from the reciprocal space to the real space, using the relation $E = \hbar^2 k^2 / 2m$:

$$\sum_k \rightarrow \left(\frac{L}{2\pi}\right)^3 \int d^3k = \int dE\, g(E). \tag{6.20}$$

Thus, we can rewrite Eq. (6.16) as

$$q = \int_0^\infty g(E) \ln\left(1 + z\, e^{-\beta E}\right) dE. \tag{6.21}$$

Inserting the density of states $g(E)$ into the above equation, performing an integration by parts[2] and considering $x = E\beta$, we obtain

$$q = \frac{2}{3} \frac{\bar{g}V}{4\pi^2} \left(\frac{2m}{\beta\hbar^2}\right)^{3/2} \int_0^\infty \frac{x^{3/2}}{z^{-1}e^x + 1} dx. \tag{6.22}$$

The above integral is well known: *PolyLogarithm* function,[3] defined as

$$Li_n(z) = \frac{1}{\Gamma(n)} \int_0^\infty \frac{x^{n-1}}{z^{-1}e^x - 1} dx, \quad \forall Re(n) > 0, \tag{6.23}$$

where $\Gamma(n)$ stands for the *Gamma* function.[3] Thus, we have

$$q = -\frac{2}{3} \frac{\bar{g}V}{4\pi^2} \left(\frac{2m}{\beta\hbar^2}\right)^{3/2} \Gamma(5/2) Li_{5/2}(-z), \tag{6.24}$$

[1] The Jacobian of this transformation is

$$dx\, dy\, dz = r^2 \sin\theta\, dr\, d\theta\, d\phi, \tag{6.18}$$

following the angles definition of Figure 3.4, and, consequently,

$$d^3k = 4\pi k^2\, dk. \tag{6.19}$$

[2] The integration by parts is based on the following relationship:

$$\int u\, dv = uv - \int v\, du,$$

where, for our particular case, we considered

$$dv = E^{1/2}\, dE$$

and

$$u = \ln\left(1 + z\, e^{-\beta E}\right).$$

[3] For further details on this function, see Appendix A.

where $\bar{g} = 2s + 1$ is the spin multiplicity. Considering electrons, it is $\bar{g} = 2$.

Finally, from the above equation, it is possible to evaluate the *grand* canonical potential. Equation (5.56) states that

$$\Phi = -\frac{1}{\beta}q \tag{6.25}$$

and then:

$$\Phi = \frac{2}{3}\frac{\bar{g}V}{4\pi^2}\left(\frac{2m}{\hbar^2}\right)^{3/2}\beta^{-5/2}\Gamma(5/2)Li_{5/2}(-z) . \tag{6.26}$$

All of the thermodynamic quantities, as discussed in Section 5.5, can now be determined.

6.3   Fermi level and chemical potential

From the *grand* canonic potential above written, it is possible to evaluate the number N of particles in the system; quantity that defines univocally the chemical potential μ. First, let us focus our attention to the zero temperature regime. For this case, the fugacity $z = e^{\mu\beta} \to \infty$ and, in accordance with the properties of the *PolyLogarithm* function (see Appendix A), we can use the following approximation:

$$Li_n(-z) \approx -\frac{\left[\ln(z)\right]^n}{n!} \quad (z \to \infty) \tag{6.27}$$

and, substituting the above equation into Eq. (6.26) we find[4]

$$\Phi = -\frac{2}{3}\frac{\bar{g}V}{4\pi^2}\left(\frac{2m}{\hbar^2}\right)^{3/2}\frac{2}{5}\mu^{5/2} \tag{6.28}$$

and, consequently, from Eq. (4.37):

$$N = -\frac{\partial\Phi}{\partial\mu} = \frac{2}{3}\frac{\bar{g}V}{4\pi^2}\left(\frac{2m}{\hbar^2}\right)^{3/2}\mu^{3/2}. \tag{6.29}$$

However, $T = 0 \Rightarrow \mu = \epsilon_F$, i.e., at zero temperature, the chemical potential defines the Fermi level. Thus, it is possible to obtain that from the above equation:

$$\epsilon_F = \left(\frac{N}{V}\frac{6\pi^2}{\bar{g}}\right)^{2/3}\frac{\hbar^2}{2m} . \tag{6.30}$$

The above result is the energy of the most energetic electron of the system.

---

[4]For this evaluation we must use some properties of the *Gamma* function, described in Appendix A.

The density of states obtained before (Eq. (6.15)), can therefore be rewritten in terms of the Fermi energy ϵ_F, and then:

$$g(E) = \frac{3}{2} \frac{N}{\epsilon_F^{3/2}} E^{1/2}. \tag{6.31}$$

The density of states at the Fermi level (of a great importance for further discussions) can also be easily written:

$$g(\epsilon_F) = \frac{3}{2} \frac{N}{\epsilon_F}. \tag{6.32}$$

Let us now consider the regime of low temperature, instead of absolute zero temperature as discussed above. For this proposal, we must consider the first correction of Eq. (6.27), that is (see Appendix A for further details):

$$Li_n(-z) \approx -\frac{[\ln(z)]^n}{n!} + 2Li_2(-1)\frac{[\ln(z)]^{n-2}}{(n-2)!}. \tag{6.33}$$

As done before for the zero temperature case, let us substitute the above approximation into Eq. (6.26). Thus, we find:

$$\Phi = -\frac{2}{3}\frac{\bar{g}V}{4\pi^2}\left(\frac{2m}{\hbar^2}\right)^{3/2}\left[\frac{2}{5}\mu^{5/2} + \frac{\pi^2}{4}\mu^{1/2}\beta^{-2}\right] \tag{6.34}$$

and, consequently,

$$N = -\frac{\partial \Phi}{\partial \mu} = \frac{2}{3}\frac{\bar{g}V}{4\pi^2}\left(\frac{2m}{\hbar^2}\right)^{3/2}\mu^{3/2}\left[1 + \frac{\pi^2}{8}\frac{1}{(\mu\beta)^2}\right]. \tag{6.35}$$

The second term in the brackets above is the temperature correction and therefore is quite smaller than the unity; thus we can do the $\mu \to \epsilon_F$ substitution (but only for the second term), and the number N of particles in the system resumes as

$$N = \frac{2}{3}\frac{\bar{g}V}{4\pi^2}\left(\frac{2m}{\hbar^2}\right)^{3/2}\mu^{3/2}\left[1 + \frac{\pi^2}{8}\frac{1}{(\epsilon_F\beta)^2}\right]. \tag{6.36}$$

Resolving now the above equation with respect to μ and also considering the expression of Fermi energy obtained in Eq. (6.30) we find:

$$\mu = \epsilon_F\left[1 + \frac{\pi^2}{8}\frac{1}{(\epsilon_F\beta)^2}\right]^{-2/3}. \tag{6.37}$$

Now we are really close to our final result, and we only need to expand [5] the previous result to therefore obtain the first correction in temperature of the chemical potential.

[5]$(1+x)^n \approx 1 + nx + \frac{1}{2}(n-1)nx^2 + \cdots \quad (x \to 0).$ \hfill (6.38)

Thus,

$$\mu \approx \epsilon_F \left[1 - \frac{\pi^2}{12} \left(\frac{k_B T}{\epsilon_F} \right)^2 \right] \tag{6.39}$$

remembering that $\beta = 1/k_B T$.

6.4 High temperature limit

For high temperature, it is possible to consider $z \to 0$ (proof of this approximation is left as an exercise), and therefore the following approximation to the *PolyLogarithm* function can be used (see Appendix A):

$$Li_n(z) \approx \sum_{k=1}^{\infty} \frac{z^k}{k^n} = z + 2^{-n} z^2 + 3^{-n} z^3 + \cdots \tag{6.40}$$

Substituting the above approximation into Eq. (6.26) we find the *grand* canonic potential:

$$\Phi = \frac{2}{3} \frac{\bar{g} V}{4\pi^2} \left(\frac{2m}{\hbar^2} \right)^{3/2} \beta^{-5/2} \Gamma(5/2)(-z). \tag{6.41}$$

From this result it is possible to evaluate any thermodynamic quantity. Thus, the total number of particles reads as

$$N = -\frac{\partial \Phi}{\partial \mu} = \frac{2}{3} \frac{\bar{g} V}{4\pi^2} \left(\frac{2m}{\hbar^2} \right)^{3/2} \beta^{-3/2} \Gamma(5/2) z. \tag{6.42}$$

Considering the fugacity $z = e^{\mu\beta}$, the chemical potential $\mu(N)$, for this high temperature limit, reads as

$$\mu(N) = \beta^{-1} \ln \left[\frac{N}{V} \frac{1}{\bar{g}} \left(\frac{4\pi \hbar^2 \beta}{2m} \right)^{3/2} \right]. \tag{6.43}$$

This result can also be achieved considering the Maxwell–Boltzmann case to Eq. (6.16), instead of the Fermi–Dirac case at high temperature (as considered), and this evaluation is left as an exercise.

Part Two

Noncooperative Magnetism

7 Diamagnetism

7.1 Localized diamagnetism

For a better understanding of the diamagnetic contribution of localized electrons, i.e., electrons in atoms, let us go back to the basis of the classical electromagnetism, more precisely to the Lenz and Faraday laws.

The British physicist Michael Faraday, among several important contributions to the science, discovered that a voltage can be induced in a coil when there is a time-dependent flux through it. As a sequence of Faraday's work, the German physicist Heinrich Lenz noticed that the current through this coil works to keep constant the flux through the coil. In other words, if the flux decreases, the magnetic field created by the coil has the same sense of the external magnetic field; while, if the flux increases, the magnetic field created by the coil has an opposite sense as that created by the external source.

The above is the physical basis behind the localized diamagnetism. Generally speaking, the electrons of the material try to shield the material to prevent external magnetic field inside itself. However, in this case, there is a difference from the Faraday-Lenz comparison above reported. For the localized diamagnetism, the induced current persists while the external magnetic field acts on the material; in a different fashion as above, where the induced current only exists when there is a magnetic field change (flux change). Finally, the magnetic field from the induced current is opposite to the external magnetic field; and, in addition, the magnetic moment associated to this induced current is the diamagnetic moment.

7.1.1 Classical formulation

In this section let us therefore obtain the magnetization and magnetic susceptibility of a localized electron, i.e., those bounded to atoms. Thus, our model starts with an electron (charge $-e$) precessing around a nucleus (charge e); and this movement is due to a Coulombian central force. This electron is therefore submitted to an electric field of the form

$$\vec{E} = \frac{e}{4\pi\epsilon_0}\frac{\vec{r}}{r^3} \tag{7.1}$$

that will be used into the Lorentz force (Eq. (1.1)). Supposing that this system is under an applied magnetic field of the form $\vec{B} = B\hat{k}$, thus, the Lorentz force $\vec{F} = -e(\vec{E} + \vec{v} \times \vec{B})$

Fundamentals of Magnetism. http://dx.doi.org/10.1016/B978-0-12-405545-2.00007-2

can be opened as follows:

$$F_x = -\frac{e^2}{4\pi\epsilon_0}\frac{x}{r^3} - ev_y B = m\frac{d}{dt}v_x,$$

$$F_y = -\frac{e^2}{4\pi\epsilon_0}\frac{y}{r^3} + ev_x B = m\frac{d}{dt}v_y. \tag{7.2}$$

Changing the above equations to polar coordinates, i.e., assuming x and y as

$$x(t) = r\cos(\omega t) \Rightarrow v_x = -r\omega\sin(\omega t) \Rightarrow dv_x/dt = -r\omega^2\cos(\omega t),$$

$$y(t) = r\sin(\omega t) \Rightarrow v_y = r\omega\cos(\omega t) \Rightarrow dv_y/dt = -r\omega^2\sin(\omega t). \tag{7.3}$$

we obtain a quadratic equation regarding to ω:

$$\omega^2 - \left(\frac{eB}{m}\right)\omega - \frac{e^2}{4\pi\epsilon_0 mr^3} = 0. \tag{7.4}$$

The solution is

$$\omega = \frac{eB}{2m} \pm \sqrt{\left(\frac{eB}{2m}\right)^2 + \frac{e^2}{4\pi\epsilon_0 mr^3}}. \tag{7.5}$$

Let us now have a break in these calculi to estimate the order of magnitude of these quantities. The electron charge and mass are, respectively, $|e| = 1.6 \times 10^{-19}$ C and $m = 9.1 \times 10^{-31}$ kg, and then $(eB/2m)^2 \approx 7.7 \times 10^{21}$ Hz^2, for $B = 1$ T (remembering that this book works within SI units and therefore those must be the units to consider). In what concerns the second term within the square root above, it is $e^2/4\pi\epsilon_0 mr^3 \approx 2.5 \times 10^{32}$ Hz^2, considering $r = 10^{-10}$ m. From these estimatives, we conclude

$$\frac{e^2}{4\pi\epsilon_0 mr^3} \gg \left(\frac{eB}{2m}\right)^2, \tag{7.6}$$

and therefore we can rewrite Eq. (7.5) as

$$\omega \approx \frac{eB}{2m} \pm \sqrt{\frac{e^2}{4\pi\epsilon_0 mr^3}} = \omega_L \pm \omega_N. \tag{7.7}$$

The first term of the above equation, i.e., $\omega_L = eB/2m$ is the Larmor frequency and it depends only on the presence of a magnetic field of induction B; while the second term, $\omega_N = \sqrt{e^2/4\pi\epsilon_0 mr^3}$, corresponds to the natural frequency of the system and is due to the Coulombian central force.

To avoid scattering of ideas, let us remember what we are looking for. We want to obtain the diamagnetic contribution of localized electrons, i.e., those bounded to atoms. For this purpose we evaluated a simple model of an electron bounded to a nucleus due to a Coulombian central force and submitted to an external magnetic field. Then we could obtain the Larmor frequency, that is proportional to the external magnetic field.

Let us now continue with our calculi. The Larmor frequency creates an extra current that, on its turn, creates an extra magnetic field in opposition to the external magnetic field (analogously to the Lenz law). This extra current is described by the electron charge times the precession frequency, i.e., the Larmor frequency, and therefore:

$$I = -e \left(\frac{\omega_{\mathrm{L}}}{2\pi} \right). \tag{7.8}$$

In accordance with Eq. (1.7), the magnetic moment due to this "electron ring" is

$$\mu = I \mathcal{A} = -\frac{e^2 B}{4m} \langle \vec{r}_\perp^2 \rangle, \tag{7.9}$$

where $\langle \vec{r}_\perp^2 \rangle$ is the quadratic mean distance from the nucleus to the electron on the plane of the orbit, i.e., $\langle \vec{r}_\perp^2 \rangle = \langle x^2 \rangle + \langle y^2 \rangle$. However, it is possible to consider, without loss of generality (i.e., the Larmor frequency keeps the same), that the magnetic field $\vec{B}$ is not aligned to the z-direction, but instead, it points along any direction. In this case, it is reasonable to consider $\langle r^2 \rangle = \langle x^2 \rangle + \langle y^2 \rangle + \langle z^2 \rangle$ instead of $\langle \vec{r}_\perp^2 \rangle$. Considering, in addition, a spherical symmetry ($\langle x^2 \rangle = \langle y^2 \rangle = \langle z^2 \rangle$), it is possible to write $\langle \vec{r}_\perp^2 \rangle = \frac{2}{3} \langle r^2 \rangle$ and therefore the magnetic moment due to the electron movement around the nucleus is

$$\mu = -\frac{e^2 B}{6m} \langle r^2 \rangle. \tag{7.10}$$

Considering the definition of magnetization ($M = N\mu$, i.e., there are N magnetic moments μ that, when summed up, hold the magnetization M), and the relationship between B and H (Eq. (1.3)), we have

$$M = -N \left(\frac{e}{3\hbar} \langle r^2 \rangle \right) \mu_0 H \mu_{\mathrm{B}}, \tag{7.11}$$

where $\mu_{\mathrm{B}} = e\hbar/2m$. The susceptibility can be evaluated (see Eq. (1.19)), being therefore:

$$\chi_d = -N \left(\frac{e}{3\hbar} \langle r^2 \rangle \right) \mu_0 \mu_{\mathrm{B}}. \tag{7.12}$$

We reached our objective: the magnetization and magnetic susceptibility due to electrons bounded to atoms. This result is know as Langevin (or Larmor) susceptibility.

Finally, it is interesting to emphasize that this susceptibility is negative and temperature independent, while the magnetization (also negative and temperature independent), only holds due to an external magnetic field H, being linear on the fourth quadrant (i.e., negative).

7.1.2   Quantum formulation

In Chapter 2 we have developed the Dirac Hamiltonian of an electron moving in a magnetic field $\vec{B} = B\hat{k}$ (Eq. (2.69)) and obtained several terms, each one corresponding to some concept. Before we go further in reading this subsection, it is strongly recommended to read that chapter. The sixth term of that equation is responsible for the diamagnetic contribution of localized electrons and can then be rewritten here:

$$\mathcal{H} = \frac{e^2}{8m} \vec{B}^2 \vec{r}_\perp^2.$$

(7.13)

Above, $\vec{r}_\perp^2$ (defined in Eq. (2.34)) is:

$$\begin{aligned}
\vec{r}_\perp^2 &= \vec{r}^2 - \frac{(\vec{r} \cdot \vec{B})^2}{\vec{B}^2} \\
&= x^2 + y^2 + z^2 - \frac{z^2 B^2}{B^2} \\
&= x^2 + y^2
\end{aligned}$$

(7.14)

and, consequently, the Hamiltonian can be rewritten as:

$$\mathcal{H} = \frac{e^2}{8m} B^2 (x^2 + y^2).$$

(7.15)

The eigenvalues of energy ϵ_n can be obtained using the Schrödinger equation and therefore:

$$\begin{aligned}
\epsilon_n &= \langle n | \mathcal{H} | n \rangle \\
&= \frac{e^2}{8m} B^2 \langle n | (x^2 + y^2) | n \rangle.
\end{aligned}$$

(7.16)

The eigenvalues of the coordinate operators are the classical coordinates and then:

$$\epsilon_n = \frac{e^2}{8m} B^2 \left(\langle x^2 \rangle + \langle y^2 \rangle \right).$$

(7.17)

We know that $\langle x^2 \rangle + \langle y^2 \rangle + \langle z^2 \rangle = \langle r^2 \rangle$ holds and, considering spherical symmetry ($\langle x^2 \rangle = \langle y^2 \rangle = \langle z^2 \rangle$), it is possible to write $\langle x^2 \rangle + \langle y^2 \rangle = \frac{2}{3} \langle r^2 \rangle$. Thus, the total energy of the system, considering N electrons, is

$$E = \frac{Ne^2}{12m} B^2 \langle r^2 \rangle.$$

(7.18)

From the thermodynamics (see Eq. (5.27)), it is possible to write the magnetization, since we know the total energy:

$$M = -\frac{\partial E}{\partial B} = -N \left(\frac{e}{3\hbar} \langle r^2 \rangle \right) \mu_0 H \mu_B,$$

(7.19)

where $\mu_B = e\hbar/2m$. The diamagnetic susceptibility then holds as

$$\chi_d = -N \left(\frac{e}{3\hbar} \langle r^2 \rangle \right) \mu_0 \mu_B.$$

(7.20)

Note that these results, developed within the quantum mechanics framework, are exactly the same as those described in the previous section (Eqs. (7.11) and (7.12)), from classical arguments.

7.2 Itinerant diamagnetism

7.2.1 Low magnetic field; low and high temperatures

In an analogous fashion as before, in the present section we will develop the magnetization and magnetic susceptibility, however, now, for itinerant electrons; and an electron
gas is a good approximation. For this purpose, we will need to recover some results
already obtained in this book. Let us start with one of the thermodynamic relationships
to the magnetization (Eq. (5.61)):

$$M = \frac{1}{\beta} \frac{\partial}{\partial B} \ln \mathcal{Z}, \tag{7.21}$$

where $\beta = 1/k_B T$ and $\mathcal{Z}$ is the *Grand* partition function.

Thus, to determine the magnetization (and, consequently, the magnetic susceptibility), it is necessary to know the logarithm of the *Grand* partition function $\mathcal{Z}$ of this
electron gas. Let us recover therefore Eq. (5.72):

$$\ln \mathcal{Z} = \sum_{k=1}^{+\infty} \ln \left(1 + z\, e^{-\beta \epsilon_k}\right), \tag{7.22}$$

where $z = e^{\beta \mu}$ stands for the fugacity and ϵ_k are the eigenvalues of the Hamiltonian. To
go further, let us discuss some points about the Hamiltonian. To describe the electron
gas, we have considered only the kinetic energy of the electrons (without either spin
or magnetic field—Chapter 6). For the present case, in addition to the kinetic energy,
we will consider the presence of a magnetic field. Later, describing the paramagnetic
behavior of an electron gas, we will consider also the spin of these electrons.

Thus, we will consider an electron gas under an applied magnetic field and to evaluate
the *Grand* partition function we need to know the eigenvalues of the Hamiltonian.
These are the Landau levels, discussed in detail in Complement 7.B (thus, we strongly
recommend to carefully read that Complement).

To go further, we need to recover the eigenvalues of energy of an electron gas
submitted to an external magnetic field $\vec{B} = B\hat{k}$ (Eq. (7.B.12)), given by the Landau
levels:

$$\epsilon_{n,k_z} = \frac{p_z^2}{2m} + 2\mu_B B \left(n + \frac{1}{2}\right), \tag{7.23}$$

that will be used into Eq. (7.22) to therefore evaluate the magnetization and magnetic
susceptibility. A similar procedure was done in Chapter 6, describing the simplest
electron gas, i.e., considering only the kinetic energy (in three dimensions). In that
case, we substitute the sum over k for an integral over also k (Eq. (6.17)), using the
concepts of Riemann's Integrals and considering large values of the wave vector $|\vec{k}|$
(i.e., large volumes in the reciprocal space) as a necessary condition. However, here,
the kinetic energy is explicitly along only the z axis and therefore the substitution from
a sum to an integral reads as

$$\sum_{k_z} \rightarrow \left(\frac{L}{2\pi}\right)\int dk_z = \left(\frac{L}{h}\right)\int dp_z, \tag{7.24}$$

considering $\vec{p} = \hbar\vec{k}$.

In addition to the kinetic energy, we also must sum over the Landau levels and therefore Eq. (7.22) must be rewritten as

$$\ln \mathcal{Z} = \frac{L}{h}\int_{-\infty}^{+\infty}\sum_{n=0}^{+\infty}\tilde{g}\ln\left\{1 + z\exp\left[-\beta\left(\frac{p_z^2}{2m} + 2\mu_B B\left(n + \frac{1}{2}\right)\right)\right]\right\}dp_z, \tag{7.25}$$

where the pre-factor L/h comes from Eq. (7.24) and

$$\tilde{g} = L^2\frac{eB}{h} \tag{7.26}$$

is the degeneracy of the Landau levels (Eq. (7.B.17))

7.2.1.1   Limit of low magnetic field ($\epsilon_F \gg \mu_B B$)

For small values of the magnetic field of induction B, the Landau levels are quite dense, since the distance between levels is directly proportional to B (see Complement 7.B for further details). For this case ($\epsilon_F \gg \mu_B B$), it is possible to change the sum over n to an integral over x, using the Euler–MacLaurin formula:

$$\sum_{n=0}^{+\infty}f\left(n + \frac{1}{2}\right) \approx \int_0^{+\infty}f(x)dx + \frac{1}{24}f'(0). \tag{7.27}$$

Using this approximation we find therefore

$$\begin{aligned}\ln\mathcal{Z} = \frac{VeB}{h^2}\int_{-\infty}^{+\infty}\int_0^{+\infty}\ln\left\{1 + z\exp\left[-\beta\left(\frac{p_z^2}{2m} + 2\mu_B Bx\right)\right]\right\}dx\,dp_z \\ -\frac{VeB}{h^2}\int_{-\infty}^{+\infty}\frac{1}{24}\left(\frac{\beta 2\mu_B B}{1 + z^{-1}\exp\left(\beta p_z^2/2m\right)}\right)dp_z.\end{aligned} \tag{7.28}$$

Let us define the first term of the above equation as $\ln\mathcal{Z}_0$. The development of the above double integral is left as an exercise to the reader and the result is

$$\ln\mathcal{Z}_0 = \frac{2\pi V(2m)^{3/2}}{h^3}\int_0^{+\infty}\epsilon^{1/2}\ln[1 + ze^{-\beta\epsilon}]d\epsilon. \tag{7.29}$$

The most interesting point of this result is that the magnetic field dependence has disappeared. Still more curious is that this result is identical to that obtained for a simple electron gas, without the presence of a magnetic field (Eqs. (6.15) and (6.21)). Thus, we can call this first term of $\ln\mathcal{Z}$, i.e., $\ln\mathcal{Z}_0$, as the zero field contribution.

Now, let us focus our attention to the second term of Eq. (7.28), that we will define as $\ln \mathcal{Z}_1$. Then:

$$\ln \mathcal{Z}_1 = -\frac{VeB}{h^2}\frac{1}{24}\int_{-\infty}^{+\infty}\frac{2\mu_B B\beta}{1-z^{-1}\exp{(\beta p_z^2/2m)}}\mathrm{d}p_z. \tag{7.30}$$

We can change the limits of the integration from $\int_{-\infty}^{+\infty}$ to $2\int_0^{+\infty}$, since it is an even function, and then the variable of integration, from p_z to $y = \beta p_z^2/2m$. Thus:

$$\ln \mathcal{Z}_1 = -\frac{V\pi\beta^{1/2}}{6h^3}(\mu_B B)^2(2m)^{3/2}\int_0^{+\infty}\frac{y^{-1/2}}{1-z^{-1}e^y}\mathrm{d}y. \tag{7.31}$$

The above integral is well known; it is the integral formulation of the *PolyLogarithm* Function[1]:

$$Li_n(z) = \frac{1}{\Gamma(n)}\int_0^\infty\frac{x^{n-1}}{z^{-1}e^x - 1}\mathrm{d}x, \quad \forall Re(n) > 0, \tag{7.32}$$

where Γ is the Gamma Function. Equation (7.31) can then be rewritten as:

$$\ln \mathcal{Z}_1 = \frac{V\pi\beta^{1/2}}{6h^3}(\mu_B B)^2(2m)^{3/2}\Gamma(1/2)Li_{1/2}(-z). \tag{7.33}$$

The logarithm of the *Grand* partition function, $\ln \mathcal{Z}$, could be separated in two contributions, $\ln \mathcal{Z}_0$ and $\ln \mathcal{Z}_1$, where the first one does not depend on the magnetic field (and we could verify that it recovers the case of a simplest electron gas, without magnetic field). Thus, from the first derivative of $\ln \mathcal{Z}$ with respect to B (proportional to the magnetization), only the second term $\ln \mathcal{Z}_1$ contributes, in such a way that it is possible to write the magnetization as:

$$\begin{aligned}
M &= \frac{1}{\beta}\frac{\partial}{\partial B}\ln \mathcal{Z} \\
&= \frac{1}{\beta}\frac{\partial}{\partial B}(\ln \mathcal{Z}_0 + \ln \mathcal{Z}_1) \\
&= \frac{1}{\beta}\frac{\partial}{\partial B}\ln \mathcal{Z}_1.
\end{aligned} \tag{7.34}$$

In addition, from Eq. (7.33), we can see that $\ln \mathcal{Z}_1 \propto B^2$ and therefore we can write

$$M = f(T)B, \tag{7.35}$$

where $f(T)$ is a function of temperature T. Thus, as expected, the magnetization is linear with magnetic field (for this limit of low values of magnetic field).

From now on, we will determine $f(T)$ for some limits (low temperature and high temperature).

---

[1] For further details on this function, see Appendix A.

7.2.1.2 Limit of low magnetic field ($\epsilon_F \gg \mu_B B$) and low temperature ($\epsilon_F \gg k_B T$)

It is important to remember that our development works only for the limit of low values of magnetic field (condition imposed to use the Euler–MacLaurin formula). To go further, we need to introduce another physical condition, that will allow us to make some mathematical approximations: low temperature limit, i.e., $\epsilon_F \gg k_B T$. In this case, the fugacity $z = e^{\mu\beta}$ tends to infinity (remember, at the low temperature limit $\mu \approx \epsilon_F$ holds, as described in Chapter 6). Thus, we have a physical condition ($\epsilon_F \gg k_B T$), that allow us to use the limit $z \to \infty$. For this situation, the *PolyLogarithm* function has a well-defined limit[2]:

$$-Li_n(-z) \approx \frac{(\ln z)^n}{n!} \quad (z \to \infty). \tag{7.36}$$

Placing above equation into Eq. (7.33), then

$$\ln \mathcal{Z}_1 = -\frac{N\beta}{4\epsilon_F}(\mu_B B)^2, \tag{7.37}$$

where (from Eq. (6.30)):

$$\epsilon_F = \frac{\hbar^2}{2m}\left(\frac{6\pi^2}{\bar{g}}\frac{N}{V}\right)^{2/3}. \tag{7.38}$$

Note that some extra properties of the Gamma function were used (see Appendix A); and, for this particular case, the magnetic field lift the degeneracy and, consequently, $\bar{g} = 1$; i.e., under a magnetic field there is only one preferential direction of the spin.

Thus, the magnetization reads as

$$M = -N\left(\frac{\mu_B}{2\epsilon_F}\right)\mu_0 H \mu_B. \tag{7.39}$$

Note that above we have used the relation $B = \mu_0 H$ (Eq. (1.3)), and therefore it means that we are within the condition of a low density gas, since this relation is valid in the vacuum. We also know that the magnetic susceptibility is given by Eq. (1.19), and then:

$$\chi_d = -N\left(\frac{\mu_B}{2\epsilon_F}\right)\mu_0 \mu_B. \tag{7.40}$$

The above result, called as Landau diamagnetic susceptibility, is therefore the magnetic susceptibility of an electron gas submitted to an external magnetic field of induction $\vec{B} = B\hat{k}$. It is also important to remember that this result is valid for *low values of magnetic field and temperature*.

---

[2]For further details on this function, see Appendix A.

It is possible to rewrite the above equations in a normalized fashion, i.e., considering $m = M/N\mu_B$, $s = \chi/(N\mu_B^2\mu_0/\varepsilon_{F0})$, $b = \mu_B B/\epsilon_{F0}$, and $t = k_B T/\epsilon_{F0}$, where $\epsilon_{F0} = \epsilon_F(\bar{g} = 2)$ is the Fermi energy at zero magnetic field (remember that a magnetic field lift the degeneracy and therefore $\bar{g} = 1$). Thus, the normalized magnetization m and normalized susceptibility s read as:

$$m = -\frac{b}{2}, \tag{7.41}$$

and

$$s = -\frac{1}{2}. \tag{7.42}$$

The behavior of the magnetization and magnetic susceptibility for these magnetic field and temperature limits is summarized in Figures 7.1a and 7.2, respectively.

7.2.1.3   Limit of low magnetic field ($\epsilon_F \gg \mu_B B$) and high temperature ($\epsilon_F \ll k_B T$)

For this limit, we must consider $z \to 0$ (see exercises of Chapter 6), and therefore it is possible to use the following approximation to the PolyLog function (Eq. (A.3))

$$Li_n(z) \approx \sum_{k=1}^{\infty} \frac{z^k}{k^n} = z + 2^{-n}z^2 + 3^{-n}z^3 + \cdots \tag{7.43}$$

Inserting the above approximation into Eq. (7.33) and considering

$$N = \frac{2}{3}\frac{\bar{g}V}{4\pi^2}\left(\frac{2m}{\hbar^2}\right)^{3/2}\beta^{-3/2}\Gamma(5/2)z, \tag{7.44}$$

(already evaluated, see Eq. (6.42)), we get

$$\ln \mathcal{Z}_1 = -\frac{N\beta^2}{6}(\mu_B B)^2, \tag{7.45}$$

and then

$$M = -N\left(\frac{\mu_B}{3k_B T}\right)\mu_0 H \mu_B. \tag{7.46}$$

Consequently,

$$\chi_d = -N\left(\frac{\mu_B}{3k_B T}\right)\mu_0 \mu_B. \tag{7.47}$$

We can also write these results in a normalized fashion, as done in Eqs. (7.41) and (7.42). Thus,

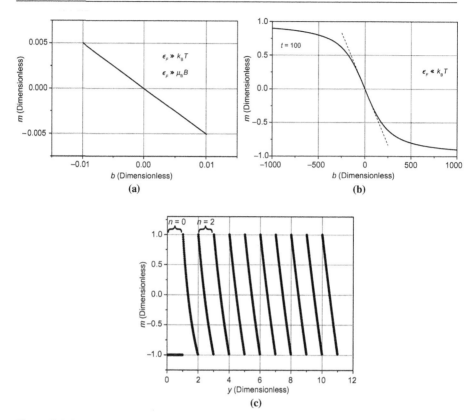

Figure 7.1 Diamagnetic contribution to the magnetization: each panel above corresponds to a certain approximation. See text for details. These are displayed via the reduced and dimensionless magnetization $m = M/N\mu_B$, magnetic field $b = \mu_B B/\epsilon_{F0}$ and temperature $t = k_B T/\epsilon_{F0}$, where $\epsilon_{F0} = \epsilon_F(\bar{g} = 2)$ is the zero field Fermi energy. (a) Low magnetic field ($\epsilon_F \gg \mu_B B$) and low temperature ($\epsilon_F \gg k_B T$) approximations. Magnetization at this limit does not depend on the temperature. See equation (7.41) for details. (b) Dashed line: low magnetic field ($\epsilon_F \gg \mu_B B$) and high temperature ($\epsilon_F \ll k_B T$) approximations. See equation (7.48) for details. Solid line: high temperature limit, without restriction on the magnetic field value. See equation (7.58) for details. These curves are for $t = 100$. (c) High magnetic field ($\epsilon_F \approx \mu_B B$) and zero temperature approximations. y is inversely proportional to the magnetic field. This oscillatory behavior of the magnetization is the de Haas-van Alphen (dHvA) effect. See Equation (7.69) for details.

$$m = -\frac{b}{3t}, \qquad (7.48)$$

and

$$s = -\frac{1}{3t}. \qquad (7.49)$$

The behavior of the magnetization and magnetic susceptibility for these magnetic field and temperature limits are summarized in Figures 7.1b and 7.2, respectively.

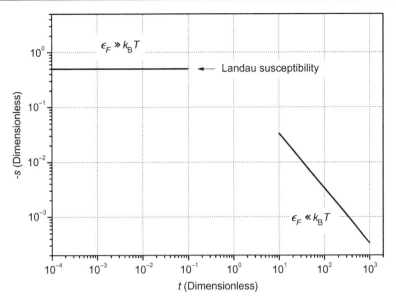

Figure 7.2 Reduced diamagnetic susceptibility for low ($\epsilon_F \gg k_B T$) and high ($\epsilon_F \ll k_B T$) temperature limits. Note that the diamagnetic susceptibility is *negative* and therefore we needed to multiply it by (-1), to then use a log–log scale. See Eqs. (7.42) and (7.49) for details.

7.2.2 High temperature; low and high magnetic field

At high temperature ($\epsilon_F \ll k_B T$), Fermi–Dirac and Boltzmann statistics merge and then, since these are equal, it is better to work with the classical one, since the logarithm of the *Grand* partition function is a sum of exponentials instead of logarithms (see Eqs. (5.72) and (5.74)). Thus, let us call Eq. (5.74):

$$\ln \mathcal{Z} = \sum_{k=1}^{+\infty} z\, e^{-\beta \epsilon_k}.$$

The eigenvalues of energy are those used before (Eq. (7.23)), and thus (analogously to Eq. (7.25)):

$$\ln \mathcal{Z} = \frac{L}{h} \int_{-\infty}^{+\infty} \sum_{k=0}^{+\infty} \bar{g} z \exp\left\{-\beta \left[\frac{p_z^2}{2m} + 2\mu_B B \left(k + \frac{1}{2}\right) \right]\right\} dp_z,$$

$$= \frac{L}{h} \bar{g} z \int_{-\infty}^{+\infty} \exp\left(-\beta \frac{p_z^2}{2m}\right) dp_z \sum_{k=0}^{+\infty} \exp\left[-\beta 2\mu_B B \left(k + \frac{1}{2}\right)\right]. \quad (7.50)$$

Then, we find[3]

$$\ln \mathcal{Z} = z \frac{V}{h^3} \left(\frac{2\pi m}{\beta}\right)^{3/2} \frac{x}{\sinh(x)}, \quad (7.51)$$

---

[3] $\int_{-\infty}^{\infty} e^{-ax^2} dx = \sqrt{\frac{\pi}{a}} \;\; Re(a) > 0, \; \sum_{k=0}^{\infty} \exp\left[-a\left(k + \frac{1}{2}\right)\right] = \frac{1}{2\sinh(a/2)}.$

where $x = \mu_B B \beta$. Note in the present case we do not use the Euler–MacLaurin formula (Eq. (7.27)), and, consequently, we are not considering the low magnetic field regime. Then, the magnetization reads as

$$M = -z \frac{V}{h^3} \left(\frac{2\pi m}{\beta} \right)^{3/2} \frac{x}{\sinh (x)} \left[\coth (x) - \frac{1}{x} \right] \mu_B. \tag{7.52}$$

The number N of particles in the system defines the chemical potential μ. For the present case ($\epsilon_F \ll k_B T$), we do not know the number of particles in the system and therefore we need to evaluate. It is possible from the following thermodynamic relation (Eq. (5.65)):

$$N = -\frac{\partial \Phi}{\partial \mu}, \tag{7.53}$$

where

$$\Phi = -\frac{1}{\beta} q = -\frac{1}{\beta} \ln \mathcal{Z}, \tag{7.54}$$

is the *Grand* canonic potential. Thus, it is easy to evaluate the number of particles, that is:

$$N = z \frac{V}{h^3} \left(\frac{2\pi m}{\beta} \right)^{3/2} \frac{x}{\sinh (x)} = \ln \mathcal{Z}. \tag{7.55}$$

Now, we just need to substitute the above equation into Eq. (7.52) and then we have

$$M = -N \left[\coth (x) - \frac{1}{x} \right] \mu_B = -N L(x) \mu_B, \tag{7.56}$$

where

$$L(x) = \coth (x) - \frac{1}{x} \tag{7.57}$$

is the Langevin function.

Similarly as before, let us rewrite the above equation in a normalized fashion; thus:

$$m = -\left[\coth \left(\frac{b}{t} \right) - \frac{t}{b} \right] = -L \left(\frac{b}{t} \right). \tag{7.58}$$

The behavior of the magnetization for this temperature limit is summarized in Figure 7.1b.

From Eq. (7.56), it is possible to recover the magnetization and magnetic susceptibility obtained for this high temperature regime (Eqs. (7.47) and (7.46)). This task is left as an exercise to the reader.

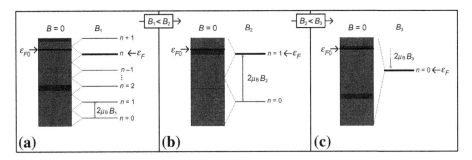

Figure 7.3 Landau levels under strong magnetic field: the de Haas–van Alphen effect.

7.2.3   High magnetic field and zero temperature: dHvA effect

Before studying the present discussion, it is strongly recommended to read Complement 7.B, where the Landau levels are discussed and explained. The de Haas–van Alphen (dHvA) effect studied here is a beautiful consequence of both, the Landau levels and the strong degeneracy of these levels when an electron gas is submitted to a strong magnetic field ($\epsilon_F \approx \mu_B B$). For the sake of clarity, we will only discuss the $T = 0$ case.

Let us start our discussion considering Figure 7.3, where each panel presents on the left the energy levels of a electrons gas closed in a box of size L, at zero temperature ($T = 0$) and zero magnetic field ($B = 0$). This point is quite important for the present discussion: our evaluation is developed considering $T = 0$, i.e., there is no statistical distribution of the electrons through the energy levels and all those are filled up to the Fermi energy $\epsilon_{F0} = \epsilon_F (\bar{g} = 2)$. The right side of each panel considers the applied field case, that increases from the left to the right, i.e., panel (a) works with a B_1 value of magnetic field, that is lower than B_2 (panel (b)), that, on its turn, is lower than B_3 (panel (c)). In other words, $B_1 < B_2 < B_3$.

Increasing magnetic field both, the degeneracy $\tilde{g} = L^2 eB/h$ and the gap ($2\mu_B B$) between the Landau levels (labeled as n) also increase. The fantastic origin of the de Haas–van Alphen effect lies exactly here, when the magnetic field increases. Let us focus our attention on Figure 7.3. The Fermi energy of the case *with* magnetic field lies in the nth Landau level, the one that contains the Fermi energy of the zero field case ϵ_{F0}, as shown in the figure. Increasing the magnetic field, the Fermi energy of the zero field case belongs then to a lower Landau level and, consequently, the Fermi level of the case with magnetic field jumps from the nth level to the $n − 1$th level, and so on down to $n = 0$. This is the de Haas–van Alphen effect, the abrupt change of the Fermi energy by changing the magnetic field; and these jumps induce oscillations on the magneto-thermal properties of the system.

Let us now understand this effect quantitatively. As mentioned above, for the sake of clarity, the zero temperature case is considered. The number N of electrons in the system (that defines the Fermi energy of the zero field case) lies between $\tilde{g}n$ and $\tilde{g}(n + 1)$, where $\tilde{g}$ is the degeneracy, i.e., the number of states for each Landau level n. Thus, the following equation holds:

$$\tilde{g}n \le N < \tilde{g}(n+1), \tag{7.59}$$

where

$$\tilde{g} = L^2 \frac{eB}{h} = \frac{B}{B_0} N = \frac{N}{y}, \tag{7.60}$$

$$B_0 = \frac{Nh}{L^2 e} \quad \text{and} \quad y = \frac{B_0}{B}. \tag{7.61}$$

Based on the definition above, Eq. (7.59) can be rewritten as

$$n \le y < n+1. \tag{7.62}$$

To go further, we need to evaluate the total energy of the system and then the magnetization, using the following thermodynamic relation:

$$M = -\frac{\partial E}{\partial B}. \tag{7.63}$$

The total energy of the completely filled Landau levels, i.e., from $n = 0$ up to $n - 1$, is the energy of the Landau levels times the number of electrons in each level (since we are working at zero temperature, the number of electrons in the level is the degeneracy). Thus:

$$E_{\{0,n-1\}} = \tilde{g} \sum_{n'=0}^{n-1} \epsilon_{n'}' = \tilde{g}\mu_B Bn^2, \tag{7.64}$$

where we have considered, from Eq. (7.B.12):

$$\epsilon_n' = 2\mu_B B \left(n + \frac{1}{2}\right). \tag{7.65}$$

The last Landau level n, partially filled and containing the Fermi energy, has the following energy:

$$E_n = [N - \tilde{g}n]\epsilon_n' = [N - \tilde{g}n]2\mu_B B \left(n + \frac{1}{2}\right). \tag{7.66}$$

Thus, the total energy is

$$E = E_{\{0,n-1\}} + E_n, \tag{7.67}$$

and then

$$\frac{E}{N} = \mu_B B \left[(2n + 1) - n(n+1)\frac{1}{y}\right]. \tag{7.68}$$

From the above equation it is possible to evaluate the magnetization of the system (Eq. (7.63)):

$$\frac{M}{N} = \left[2n(n+1)\frac{1}{y} - (2n+1) \right] \mu_B, \quad n \leq y < n+1 \,. \tag{7.69}$$

This magnetization has an oscillatory behavior of unitary period on y and peak-to-peak amplitude of $2\mu_B$, as presented in Figure 7.1c. The amplitude of these oscillations is further discussed in the exercises.

Complements

7.A   Experimental example

As discussed before, all of the materials have a diamagnetic contribution, even those extremely magnetic; and, for these cases, the diamagnetic contribution is much smaller than the other contributions. On the other hand, there are some materials with only a diamagnetic contribution and, consequently, quite evident. Examples are: water, gold, silver, most of the organic materials, and much more.

Figure 7.4 presents the magnetization as a function of magnetic field for the sample holder (plastic straw and Teflon) of a commercial magnetometer. Note the magnetization is linear and negative; and, in addition, does not depend on the temperature—as predicted by the models presented in this chapter.

7.B   Landau levels

The ideas and results described in the present Complement are of extreme importance to Section 7.2, where we studied the diamagnetic behavior of an ideal electron gas under an applied magnetic field. Thus, in the present Complement we will study and analyze the energy levels of an electron with moment $\vec{p}$ moving in a box of length L and under a magnetic field pointing along the z axis, i.e., $\vec{B} = B\hat{k}$.

To start, let us recover the results from Chapter 2, more precisely, Eq. (2.21):

$$\mathcal{H} = \frac{1}{2m} \left(\vec{p} + e\vec{A} \right)^2, \tag{7.B.1}$$

where $\vec{A}$ is the vector potential.[4] For the case $\vec{B} = B\hat{k}$, we can choose the Landau gauge (see Eq. (2.23) for further details):

$$\vec{A} = Bx\hat{j}. \tag{7.B.2}$$

Thus, we can rewrite the above Hamiltonian as:

$$\mathcal{H} = \frac{1}{2m} \left(p_x^2 + \left(p_y + eBx \right)^2 + p_z^2 \right). \tag{7.B.3}$$

---

[4] $\vec{B} = \vec{\nabla} \times \vec{A}$.

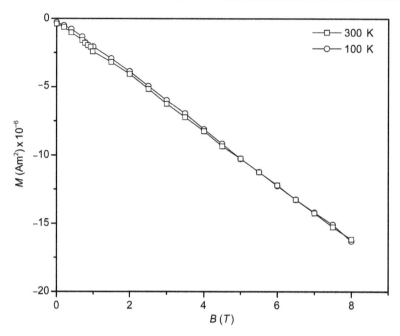

Figure 7.4 Experimental diamagnetic contribution of a sample holder (plastic straw and Teflon) of a commercial magnetometer. Measurement done by Prof. Angelo Gomes, UFRJ-Brazil.

Now, we can use this Hamiltonian into the time-independent Schrödinger equation:

$$\mathcal{H}\Psi = \epsilon\Psi, \tag{7.B.4}$$

where the total wave function can be written as

$$\Psi(x, y, z) = \phi_n(x)\phi_{n_y}(y)\phi_{n_z}(z). \tag{7.B.5}$$

Above, n, n_y, and n_z are quantum numbers associated to the wave functions.

Well, we know from Chapter 2 that the linear momentum operator is $\vec{p} = -i\hbar\vec{\nabla}$ and, for the case where we have only kinetic energy $p^2/2m$, Schrödinger equation has as solution:

(i) $(\hbar k)^2/2m$ as energy eigenvalues and,

(ii) $\Psi = e^{i\vec{k}\cdot\vec{r}}$ (plane wave) as wave function.

However, the Hamiltonian of Eq. (7.B.3) has a x coordinate variable (from Eq. (7.B.2)), and therefore we cannot propose a plane wave solution to this coordinate. On the other hand, we can consider plane waves to the other two coordinates y and z; and the total wave function reads as:

$$\Psi(x, y, z) = \phi_n(x)e^{ik_y y} e^{ik_z z}. \tag{7.B.6}$$

Equation (7.B.3) can then be rewritten as:

$$\frac{1}{2m}\left[-\hbar^2\frac{d^2}{dx^2}+\left(\hbar k_y+eBx\right)^2+\hbar^2 k_z^2\right]\phi_n(x)=\epsilon_n\phi_n(x),\tag{7.B.7}$$

or

$$\left[-\frac{\hbar^2}{2m}\frac{d^2}{dx^2}+\frac{1}{2}m\omega^2\left(x+x_0\right)^2\right]\phi_n(x)=\epsilon'_n\phi_n(x),\tag{7.B.8}$$

where

$$x_0=\frac{\hbar k_y}{eB},\tag{7.B.9}$$

$$\omega=\frac{eB}{m},\tag{7.B.10}$$

$$\epsilon'_n=\epsilon_n-\frac{\hbar^2 k_z^2}{2m}.\tag{7.B.11}$$

Note Eq. (7.B.8) is the equation of an harmonic oscillator centered at x_0, and the oscillation frequency ω is the Cyclotron Frequency.[5]

The eigenvalues of the quantum harmonic oscillator are

$$\boxed{\epsilon'_n=\left(n+\frac{1}{2}\right)\hbar\omega}\,,\tag{7.B.12}$$

and these are the Landau levels of energy! See Figure 7.5.

For the sake of clarity, it is important to pause our development to remember our ideas. We have a free electron moving under a magnetic field $\vec{B}=B\hat{k}$. Solving Schrödinger equation we could propose a plane wave to y and z directions, while for the x coordinate we found a harmonic oscillator centered at $x_0=\hbar k_y/eB$, with wave function $\phi_n(x)$ and energy eigenvalues ϵ'_n. Let us now go back to our development.

Each quantum state n has several harmonic oscillators, each one centered at x_0 and all of them with the same frequency ω. However, where are these centers? Obviously, these centers must be inside the physical limits of the box in which the electrons are into; and, mathematically speaking, we can write:

$$0\le\left(x_0=\frac{\hbar k_y}{eB}\right)\le L.\tag{7.B.13}$$

However, which values can k_y assume? Considering the Born–von Karman boundary conditions (see Eq. (6.6)), i.e., $\phi_{n_y}(y+L)=\phi_{n_y}(y)$, we easily find that these plane waves must satisfy the condition $e^{ik_y L}=1$, and, consequently, we can write

$$k_y=\frac{2\pi}{L}n_y,\tag{7.B.14}$$

---

[5]Before, we have obtained the precession frequency (Eq. (7.7)) of an electron bounded to a nucleus (subjected therefore to a Coulombian potential and a magnetic field), and then derived the Larmor frequency $\omega_L=eB/2m$. It is easy to verify from Eq. (7.5) that, if we make zero the Coulombian contribution, we can thus recover the cyclotron frequency $\omega=eB/m$.

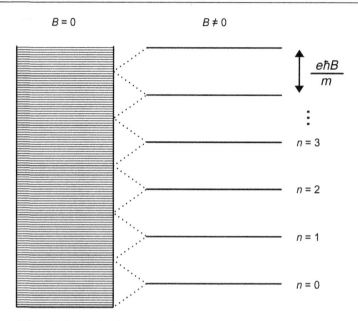

Figure 7.5 Left: Energy levels scheme of an electron in a box. Right: Landau levels.

where $n_y = 0, 1, \ldots$ is the quantum number associated to the y wave function, i.e., $\phi_{n_y}(y)$. Substituting Eq. (7.B.14) into Eq. (7.B.13) we have

$$0 \leq n_y \leq L^2 \frac{eB}{h}. \tag{7.B.15}$$

Thus, for a certain n there are several oscillators centered at

$$x_0 = \frac{h}{eBL} n_y, \tag{7.B.16}$$

and the position of these centers are also quantized, due to its dependence with n_y. In addition, the maximum value that n_y can assume is

$$\tilde{g} = L^2 \frac{eB}{h}. \tag{7.B.17}$$

Summarizing, for a unique value of energy ϵ'_n of the oscillator, we may find n_y wave functions $\phi_{n_y}(y)$ without changes in the energy of this oscillator. Thus, for each n there are n_y possibilities, being therefore $\tilde{g}$ the degeneracy of the state n. Finally, we can also emphasize another aspect: the degeneracy $\tilde{g}$ depends on the magnetic field B, in such a way that as bigger the magnetic field is, bigger is the degeneracy, i.e., bigger is the number of possible centers x_0 for the same n. Also note that, for the zero magnetic field case, the dependency of x in Eq. (7.B.7) disappears and then we recover the equation of a particle closed in a box, with a plane wave for the three directions x, y, and z.

8 Paramagnetism

8.1 Localized paramagnetism

In the present section we will deal with an important type of magnetic arrangement: the paramagnetic behavior of localized magnetic moments, i.e., those bonded to ions. First, a general description of important thermodynamic quantities will be given and then two cases will be extensively discussed: the classical case, where the projection of the angular momentum along the magnetic field vector can be considered as a continuous variable; and the quantum case, that, on its turn, the projection is quantized.

8.1.1 Thermodynamic quantities

To start this subsection, let us consider the partition function, that can be a function of temperature T, magnetic field of induction B, and the number of particles, that, for the sake of simplicity, we can consider $N = 1$, i.e., the partition function of a single particle.

$$Z(T, B, 1) = \sum_i \exp(-\beta\varepsilon_i), \tag{8.1}$$

where ε_i is the energy spectra of the system considered. However, generally speaking, the magnetic energy is

$$\varepsilon = -\vec{\mu} \cdot \vec{B}, \tag{8.2}$$

where the magnetic moment μ is written in μ_B units. We can also consider, without losing our aim of a general description, that the magnetic energy and thermal energy always appear together in a fraction

$$x = \frac{b}{t} = \frac{M_s B}{k_B T} = \beta b, \tag{8.3}$$

where M_s is the saturation value of the magnetization, or, in other words, the maximum projection of the magnetic moment along the magnetic field vector. Thus, we are able to consider, based on the assumptions done above, that the partition function for a single particle can be rewritten as

$$Z(T, B, 1) \rightarrow Z(x, 1) \equiv Z(x). \tag{8.4}$$

Fundamentals of Magnetism. http://dx.doi.org/10.1016/B978-0-12-405545-2.00008-4

Finally, as described in Chapter 5, for an ensemble of N distinguishable particles, we have

$$Z(x, N) = [Z(x)]^N. \tag{8.5}$$

From now on, based on the above development, we will derive the general description of some thermodynamic quantities, starting from the Helmholtz free energy.

8.1.1.1   Helmholtz free energy

From Eq. (5.37) we know that

$$F(x, N) = Nf(x), \tag{8.6}$$

where $f(x)$ above defined is the reduced (one-particle) Helmholtz free energy and therefore

$$f(x) = -\frac{1}{\beta} \ln [Z(x)]. \tag{8.7}$$

8.1.1.2   Magnetization

From Eqs. (5.39) and (5.40) we know that the magnetization depends on the Helmholtz free energy $F(x, N)$

$$M(x, N) = -\frac{\partial}{\partial B} F(x, N), \tag{8.8}$$

$$= N M_s m(x), \tag{8.9}$$

where the above defined $m(x)$ is the reduced and normalized magnetization, since it ranges from $+1$ down to -1. One more single step evaluating the equation leads to

$$N M_s m(x) = -\frac{\partial}{\partial b} [Nf(x)] \frac{\partial b}{\partial B} \tag{8.10}$$

and finally we obtain the normalized magnetization in terms of reduced Helmholtz free energy $f(x)$

$$m(x) = -\frac{\partial f(x)}{\partial b}, \tag{8.11}$$

$$= -\frac{\partial}{\partial b} \left\{ -\frac{1}{\beta} \ln [Z(x)] \right\}. \tag{8.12}$$

The final relationship for the magnetization as a function of the partition function of a single particle is

$$m(x) = \frac{\partial}{\partial x} \{ \ln [Z(x)] \}. \tag{8.13}$$

8.1.1.3   Magnetic entropy

In a similar fashion as before, we can derive the reduced entropy as a function of the partition function of a single particle. Thus, let us start from the relationship of the entropy in terms of the Helmholtz free energy $F(x, N)$, already discussed before and presented in Eq. (5.36).

$$S(x, N) = k_B \beta^2 \frac{\partial}{\partial \beta} F(x, N), \tag{8.14}$$

$$= N k_B s(x), \tag{8.15}$$

where the above defined $s(x)$ is the reduced entropy. For a while, it is not possible to predict its upper and lower limits, but these will be discussed carefully in the next subsection. Thus, the reduced entropy reads as

$$s(x) = \beta^2 \frac{\partial f(x)}{\partial \beta}, \tag{8.16}$$

$$= \beta^2 \frac{\partial}{\partial \beta} \left\{ -\frac{1}{\beta} \ln \left[Z(x) \right] \right\} \tag{8.17}$$

and therefore we find a closed expression, in terms of the partition function of a single particle

$$\boxed{s(x) = \ln \left[Z(x) \right] - x m(x)} . \tag{8.18}$$

The above result can also be found from the thermodynamic definition of the Helmholtz free energy

$$F = U - TS \implies f = u - \frac{s}{\beta}, \tag{8.19}$$

where the above reduced expression on the right was obtained considering the reduced internal energy $u(x)$, defined as

$$U(x, N) = N u(x) = -\vec{M} \cdot \vec{B} \implies u(x) = -m(x) b. \tag{8.20}$$

Considering $f(x)$ and $u(x)$ above described (Eqs. (8.7) and (8.20), respectively) into the reduced thermodynamic relation $f = u - s/\beta$, we recover therefore the reduced entropy $s(x)$ presented in Eq. (8.18).

8.1.1.4   Specific heat

Finally, let us deal with the specific heat, a thermodynamic quantity of extreme importance for condensed matter problems. However, following the scope of this book, here we will only describe the magnetic contribution to the total specific heat. Following the same idea as before, we start with the thermodynamic relation for the specific heat at constant magnetic field of induction B (from Eq. (4.47))

$$C_B(x, N) = -\beta \frac{\partial}{\partial \beta} S(x, N) \bigg|_B , \tag{8.21}$$

$$= N k_B c_b(x), \tag{8.22}$$

where $c_b(x)$ is the reduced (one-particle) specific heat. Again, its limit cannot be predicted for a while, but will be extensively discussed further in this chapter, for the classical and quantum cases.

From the above equation, and considering the reduced entropy $s(x)$, it is possible to write

$$c_b(x) = -\beta \frac{\partial}{\partial \beta} \left\{ \ln \left[Z(x) \right] - xm(x) \right\} \Big|_b$$

$$= -\beta \left\{ \frac{\partial}{\partial x} \ln \left[Z(x) \right] - m(x) - x \frac{\partial m(x)}{\partial x} \right\} \frac{\partial x}{\partial \beta} \Big|_b \qquad (8.23)$$

and finally

$$c_b(x) = x^2 \frac{\partial m(x)}{\partial x}. \qquad (8.24)$$

Thus, we found the specific heat as a function of the magnetization, that depends, on its turn, on the partition function of a single particle.

There is also the specific heat at constant magnetization C_M (for further details, see Chapter 5), and, analogously to the above description, it is possible to write

$$c_m(x) = -\beta \frac{\partial s(x)}{\partial \beta} \Big|_m, \qquad (8.25)$$

where

$$C_M(x, N) = N k_B c_m(x). \qquad (8.26)$$

Again, considering the reduced entropy $s(x)$ we find

$$c_m(x) = -\beta \frac{\partial}{\partial \beta} \left\{ \ln \left[Z(x) \right] - xm(x) \right\} \Big|_m$$

$$= -\beta \left\{ \frac{\partial}{\partial x} \ln \left[Z(x) \right] - m(x) \right\} \frac{\partial x}{\partial \beta}, \qquad (8.27)$$

however, we know the normalized magnetization $m(x)$ (Eq. (8.13)) and therefore we find an interesting result

$$c_m = 0. \qquad (8.28)$$

Is it expected? This answer is left as an exercise to the reader.

From now on, let us work on two specific cases: a classical one, in which the magnetic moment can assume any projection along the magnetic field vector; and the quantum case, in which the projection is quantized.

8.1.2   Classical and quantum cases

8.1.2.1   Partition function

As discussed in Chapter 5, if we know the partition function, we can therefore derive all of the thermodynamic quantities. Thus, let us focus our attention to the partition function of a single particle

$$Z(T, B, 1) = \sum_i \exp(-\beta \varepsilon_i), \tag{8.29}$$

where ε_i is the energy spectra of a single particle. Back to the basis, the Zeeman magnetic energy is

$$\varepsilon = -\vec{\mu} \cdot \vec{B}, \tag{8.30}$$

where the magnetic moment μ is written in μ_B units. To simplify the problem, we will consider $\vec{B}$ along the z-axis, i.e., $\vec{B} = B\hat{k}$.

Classical case: For this case, the projection of the spin along the magnetic field of induction $\vec{B} = B\hat{k}$ can assume continuous values. Considering θ the angle between the magnetic moment $\vec{\mu}$ and the z-axis, the magnetic energy reads as

$$\varepsilon = -\mu \mu_B B \cos(\theta). \tag{8.31}$$

Figure 3.4 clarifies this scenario. Obviously, ε is equivalent to ε_i and we also must change the sum to an integral

$$\sum_i \rightarrow \int d\Omega_{sa}, \tag{8.32}$$

where $d\Omega_{sa}$ is an infinitesimal solid angle. It can be written in terms of the zenithal $0 \leq \theta < \pi$ and azimuthal $0 \leq \phi < 2\pi$ angles

$$d\Omega_{sa} = \sin(\theta)d\theta \, d\phi. \tag{8.33}$$

Thus, the partition function for a single particle reads as

$$Z(T, B, 1) = Z(x) = \int_0^{2\pi} \int_0^{\pi} \exp\left[\beta \mu \mu_B B \cos(\theta)\right] \sin(\theta)d\theta \, d\phi$$

$$= 2\pi \int_{-1}^{1} \exp(xy)dy, \tag{8.34}$$

where above, we changed the variables in accordance with $y = \cos(\theta)$ and

$$x = \frac{b}{t} = \frac{\mu \mu_B B}{k_B T} = \beta \mu \mu_B B. \tag{8.35}$$

Finally, the canonical partition function for a single classical particle reads as

$$Z(x) = 4\pi \frac{\sinh(x)}{x}.$$
(8.36)

Quantum case: For the quantum case, the magnetic energy, described in detail in Chapter 3 (Eq. (3.53)), reads as

$$\varepsilon_i \to \varepsilon_{m_j} = g m_j \mu_B B,$$
(8.37)

where m_j are the eigenvalues of the J_z operator and range from $+j$ down to $-j$, in unitary steps. Thus, we can write the partition function

$$Z(T, B, 1) = Z(x) = \sum_{m_j=-j}^{+j} \exp(-\beta \varepsilon_{m_j})$$

$$= \sum_{m_j=-j}^{+j} \exp\left(-\frac{x}{j} m_j\right)$$

$$= \sum_{m_j=-j}^{+j} v^{m_j},$$
(8.38)

where $v = \exp(-x/j)$ and

$$x = \frac{b}{t} = \frac{g j \mu_B B}{k_B T} = \beta g j \mu_B B.$$
(8.39)

The sum above has a closed solution[1] and therefore

$$Z(x) = \frac{\sinh(ax)}{\sinh(bx)},$$
(8.41)

where

$$a = 1 + \frac{1}{2j} \quad \text{and} \quad b = \frac{1}{2j}.$$
(8.42)

From the results derived above we can therefore obtain the thermodynamic quantities.

[1]
$$\sum_{n=-l}^{+l} x^n = x^{-l} \frac{[-1 + x^{(1+2l)}]}{[x-1]}.$$
(8.40)

8.1.2.2 Magnetization

This is one of the most important quantities for condensed matter problems and, of course, the core of this book. From Eq. (8.13) we know that

$$m(x) = \frac{\partial}{\partial x} \left\{ \ln \left[Z(x) \right] \right\}$$

(8.43)

and therefore it is easy to evaluate the magnetization for both, classical and quantum cases.

Classical case: Considering

$$\frac{\partial}{\partial x} \left\{ \ln \left[4\pi \frac{\sinh (x)}{x} \right] \right\} = \frac{x}{\sinh (x)} \frac{\partial}{\partial x} \left[\frac{\sinh (x)}{x} \right],$$

(8.44)

$$= \coth (x) - \frac{1}{x}$$

(8.45)

then the magnetization reads as

$$m(x) = L(x) = \coth (x) - \frac{1}{x}.$$

(8.46)

The above equation is the Langevin function and can only assume values from +1 down to −1 (remember, it is the normalized magnetization). On the other hand, we know that the absolute value of the magnetization is $M(T, B, N) = N M_s m(x)$; however, before going further, we need to know M_s, i.e., the saturation value of the magnetization. In other words, what is the maximum projection of the magnetic moment along the magnetic field direction (in our case, the z-axis, since we considered $\vec{B} = B\hat{k}$)? Intuitively, it is the value of the total magnetic moment, i.e., $\mu \mu_B$. Indeed, it can be confirmed comparing Eqs. (8.3) and (8.35). Thus, the absolute value of the classical magnetization is

$$M(T, B, N) = N\mu\mu_B L(x).$$

(8.47)

Quantum case: In a similar fashion as before, we need to evaluate

$$\frac{\partial}{\partial x} \left\{ \ln \left[\frac{\sinh (ax)}{\sinh (bx)} \right] \right\} = \frac{\sinh (bx)}{\sinh (ax)} \frac{\partial}{\partial x} \left[\frac{\sinh (ax)}{\sinh (bx)} \right],$$

(8.48)

$$= a \coth (ax) - b \coth (bx)$$

(8.49)

and then obtain the normalized magnetization for this quantum case:

$$m(x) = B_j(x) = a \coth (ax) - b \coth (bx),$$

(8.50)

where the above equation is the well-known Brillouin function.

Again, the absolute value of the magnetization is $M(T, B, N) = N M_s m(x)$ and we need to know the saturation value of the magnetization M_s. The maximum projection of this quantum spin along the z axis is $gj\mu_B$ and therefore it is M_s (that can also be obtained comparing Eqs. (8.3) and (8.39)). Thus,

$$M(T, B, N) = Ngj\mu_B B_j(x). \tag{8.51}$$

We are now discussing the quantum case, i.e., the projection of the total magnetic moment along the magnetic field vector. However, if the total magnetic moment is quite huge, there are infinity projections and therefore the system behaves as a classical one. Thus, the Brillouin function recovers the Langevin function after the $j \rightarrow \infty$ limit. Considering the expansion of the hyperbolic cotangent function for low values of x

$$\coth(x) \xrightarrow{x \rightarrow 0} \frac{1}{x} + \frac{x}{3} - \frac{x^3}{45} + \mathcal{O}(x^5), \tag{8.52}$$

it is easy to obtain

$$B_j(x) \xrightarrow{j \rightarrow \infty} \coth(x) - \frac{1}{x} = L(x). \tag{8.53}$$

Details of this evaluation are left as an exercise to the reader.

For both cases, increasing x (for instance, increasing the magnetic field b for a constant temperature t) the ensemble of spins increases the projection along the magnetic field, increasing therefore the magnetization. Note that $x = 0$ implies in $m = 0$, i.e., if there is no magnetic field (or infinity temperature) each spin points to one direction and the sum is zero. The same idea can be seen from the magnetization as a function of $x^{-1} = t/b$ (for instance, changing temperature t for a constant magnetic field b). For $x^{-1} = 0$, i.e., fully ordered system (either zero temperature or infinity magnetic field), the magnetization is $m = 1$ (all of the spins point along the magnetic field), and then decreases to zero increasing x^{-1}, because the thermal energy destroys the arrangement. Thus, Figure 8.1 presents the magnetization for these two cases: (a) as a function of $x = b/t$ (understood as an action of the magnetic field) and (b) as a function of $x^{-1} = t/b$ (action of the temperature).

From now on, let us explore the limits of these results, when x tends to zero and infinity.

$\Longrightarrow$ *Limit* of low magnetic field and high temperature ($b \ll t \Rightarrow x \rightarrow 0$).

To evaluate this limit for both, the classical and quantum cases, we need to consider the expansion of the hyperbolic cotangent function for small values of x, given in Eq. (8.52) and then substitute into the Langevin and Brillouin functions.

Classical case: The Langevin function in this limit reads therefore as

$$L(x) \xrightarrow{x \rightarrow 0} \frac{1}{x} + \frac{x}{3} - \frac{x^3}{45} - \frac{1}{x} = \frac{x}{3} - \frac{x^3}{45}. \tag{8.54}$$

Thus, for low values of x

$$m(x) = \frac{x}{3} - \frac{x^3}{45}. \tag{8.55}$$

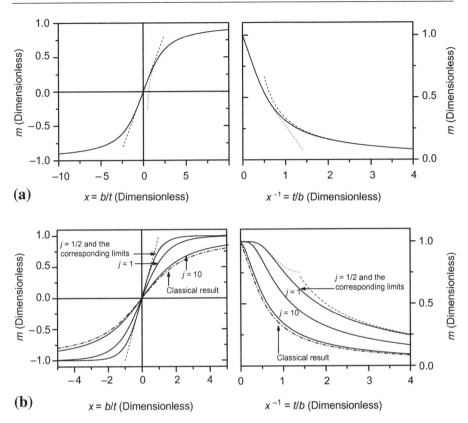

Figure 8.1 (Left panels) Magnetization m as a function of $x = b/t$. These figures can be seen as the influence of the magnetic field on the magnetization for a constant temperature. (Right panels) Magnetization m as a function of $x^{-1} = t/b$. These figures can be seen as the influence of the temperature on the magnetization, for a constant magnetic field. Note (i) the limit for low magnetic field and high temperature ($x \to 0$ and $x^{-1} \to \infty$—dashed line), in which we only consider the first-order correction (in this case the term proportional to x) and (ii) the limit for high magnetic field and low temperature ($x \to \infty$ and $x^{-1} \to 0$—dotted line). See text for details. (a) Classical case. (b) Quantum case—the classical result (dash-dotted line) is also presented, for the sake of comparison.

and, consequently,

$$M(T, H, N) = \frac{N(\mu\mu_{\mathrm{B}})^2 \mu_0 H}{3k_{\mathrm{B}}T}. \qquad (8.56)$$

The magnetic susceptibility can then be obtained and reads as

$$\chi = \frac{N\mu_0(\mu\mu_{\mathrm{B}})^2}{3k_{\mathrm{B}}T}. \qquad (8.57)$$

Quantum case: For the quantum case, the total angular momentum j plays an important rule. It ranges from 1/2 up to infinity and must be taken into account when evaluating the limit of the thermodynamic quantity. Since j is at the denominator of the argument of the hyperbolic cotangent function, there are two possibilities: (i) it is huge and then ratifies the $x \to 0$ limit or (ii) it is 1/2 (minimum value) and then does not change the $x \to 0$ limit. In other words, this limit is valid for any value of j. Thus, the Brillouin function becomes

$$B_j(x) \xrightarrow{x \to 0} a \left[\frac{1}{ax} + \frac{ax}{3} - \frac{(ax)^3}{45} \right] - b \left[\frac{1}{bx} + \frac{bx}{3} - \frac{(bx)^3}{45} \right]$$

$$= \frac{x}{3}(a^2 - b^2) - \frac{x^3}{45}(a^4 - b^4) \tag{8.58}$$

and finally, for $x \to 0$, we have

$$m(x) = \frac{x}{3}(a^2 - b^2) - \frac{x^3}{45}(a^4 - b^4) \tag{8.59}$$

and, consequently,

$$M(T, H, N) = \frac{Ng^2 j(j+1)\mu_B^2}{3k_B T} \mu_0 H. \tag{8.60}$$

The magnetic susceptibility can also be evaluated for this case, and we then find the famous Curie Law

$$\chi = \frac{C}{T}, \tag{8.61}$$

where

$$C = \frac{N\mu_0 p_{\text{eff}}^2}{3k_B} \tag{8.62}$$

is the Curie constant and

$$p_{\text{eff}}^2 = g^2 j(j+1)\mu_B^2, \tag{8.63}$$

the paramagnetic effective moment. These last equations are of great importance to the magnetism. As we will discuss soon (see complements of this chapter), when we measure the magnetic susceptibility of a certain material, the behavior at high temperature (the present limit) gives, from the fitting of a straight line on the temperature dependence of the inverse susceptibility data, the paramagnetic effective moment and therefore information concerning the effective total angular momentum of the magnetically isolated ions.

$\implies$ *Limit* of high magnetic field and low temperature ($b \gg t \Rightarrow x \to \infty$).

In an analogous fashion as before, the expansion of the hyperbolic cotangent function is needed; however, now, for high values of x

$$\coth(x) \xrightarrow{x \to \infty} 1 + 2e^{-2x}. \tag{8.64}$$

Classical case: The magnetization for this $x \to \infty$ limit reads therefore as

$$m = L(x) = 1 + 2e^{-2x} - \frac{1}{x}. \tag{8.65}$$

Quantum case: For this case, it is important to emphasize one point. We are dealing with high values of the argument of the hyperbolic cotangent function and therefore x, in addition to be high, must satisfy $x \gg 2j$; otherwise, the argument of the hyperbolic cotangent function can be not high enough to satisfy the expansion. In other words, for the quantum case ($x \to \infty$ and $x \gg 2j$), the Brillouin function reads as

$$B_j(x) \xrightarrow{x \to \infty} a\left[1 + 2e^{-2ax}\right] - b\left[1 + 2e^{-2bx}\right] \tag{8.66}$$

and finally

$$m(x) = 1 + 2\left[ae^{-2ax} - be^{-2bx}\right]. \tag{8.67}$$

Note therefore this result is intrinsically quantum, i.e., cannot be obtained from classical approaches, since now we cannot assume $j \to \infty$.

Both expansions, for low and high values of x, are represented in Figure 8.1.

8.1.2.3 Magnetic entropy

As described before, the magnetic contribution to the entropy has the general form

$$s(x) = \ln\left[Z(x)\right] - xm(x) \tag{8.68}$$

and therefore to obtain this quantity, like the other quantities, we only need to know the partition function of a single particle.

Classical case: Considering the partition function and magnetization already described above (Eqs. (8.36) and (8.46)), it is simple to obtain the magnetic entropy

$$s(x) = \ln\left[4\pi \frac{\sinh(x)}{x}\right] - xL(x). \tag{8.69}$$

Quantum case: Analogously to the quantum case, considering Eqs. (8.41) and (8.50)

$$s(x) = \ln\left[\frac{\sinh(ax)}{\sinh(bx)}\right] - xB_j(x). \tag{8.70}$$

What do we expect for the behavior of the entropy as a function of x? Roughly speaking, the entropy is a measure of the disorder in the system and therefore, for low values of x, i.e., low magnetic field and high temperature, it must tend to a constant value, because the spins are fully disordered. The entropy then goes to the logarithm of the number of accessible states Ω. On the other hand, when x tends to high values, i.e., high magnetic field and low temperature, the entropy must then go to zero. Why? First, we must recall the third law of the thermodynamics, that states that entropy goes to zero when the temperature goes to zero. Second, in this limit, the spins point along only one direction (along the magnetic field), and therefore only one state is possible to be accessed. Thus, we have $\ln(1) = 0$. Finally, we expect a monotonically increasing entropy curve as a function of x, starting from zero and saturating at the logarithm of the number of accessible states Ω.

Figure 8.2 presents the normalized and reduced entropy $\underline{s}(x) = s(x)/\ln(\Omega)$: (a) as a function of $x = b/t$ (understood as an action of the magnetic field) and (b) as a function of $x^{-1} = t/b$ (action of the temperature). Note Ω is the number of accessible states for each case (it is 4π for the classical case and $(2j + 1)$ for the quantum case—see discussion below to understand).

The discussion started above will be clearer after we evaluate the limits of the entropy, as done below.

$\Longrightarrow$ *Limit* of low magnetic field and high temperature ($b \ll t \Rightarrow x \to 0$).

For the entropy we need to know the expansion of the hyperbolic sine function; and for this limit it reads as

$$\sinh(x) \xrightarrow{x \to 0} x + \frac{x^3}{6} + \frac{x^5}{120} + \mathcal{O}(x^7). \tag{8.71}$$

Classical case: For this case, considering also the expansion for the Langevin function (Eq. (8.55)), we have

$$s(x) \xrightarrow{x \to 0} \ln(4\pi) + \ln\left(\frac{x + x^3/6}{x}\right) - x\left(\frac{x}{3}\right). \tag{8.72}$$

To go further, the expansion below is needed

$$\ln(1 + x^2) \xrightarrow{x \to 0} x^2 - \frac{x^4}{2} + \mathcal{O}(x^6) \tag{8.73}$$

and therefore

$$s(x) = \ln(4\pi) - \frac{x^2}{6}. \tag{8.74}$$

Obviously, there are corrections for higher orders of x and it is not straightforward to be obtained as the x^2 correction above. In this sense, here we present only the final result[2] that must be added to the above equation

$$+\frac{x^4}{60}. \tag{8.75}$$

[2]It can be easily obtained from a computer program such as Mathematica, Maple, or MatLab.

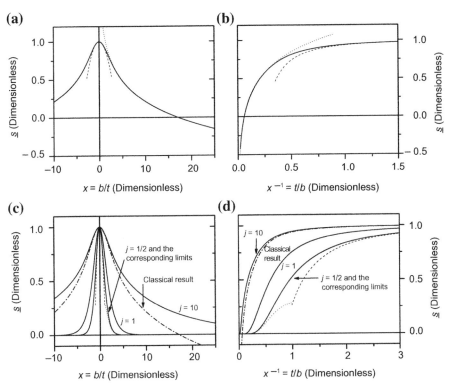

Figure 8.2 (a, c) Reduced and normalized magnetic entropy $\underline{s}(x) = s(x)/\ln(\Omega)$ as a function of $x = b/t$. These figures can be seen as the influence of the magnetic field on the entropy for a constant temperature. (b, d) Reduced and normalized magnetic entropy $\underline{s}(x) = s(x)/\ln(\Omega)$ as a function of $x^{-1} = t/b$. These figures can be seen as the influence of the temperature on the entropy, for a constant magnetic field. Note (i) the limit for low magnetic field and high temperature ($x \to 0$ and $x^{-1} \to \infty$—dashed line), in which we only consider the first-order correction (in this case, the term proportional to x^2) and (ii) the limit for high magnetic field and low temperature ($x \to \infty$ and $x^{-1} \to 0$—dotted line). See text for details. (a) Classical case. (b) Quantum case—the classical result (dash-dotted line) is also presented, for the sake of comparison.

We can understand the first term of Eq. (8.74) as follows: as discussed above, we know that for this limit all accessible states can be occupied, since the thermal energy allows the spins to point along any direction. Thus, we are dealing with an equiprobable case, and then the entropy is $\ln(\Omega)$, where Ω is the number of accessible states and reads as

$$\Omega = \Omega_{sa} = \int d\Omega_{sa} = \int_0^{2\pi} \int_0^{\pi} \sin(\theta)d\theta \, d\phi = 4\pi, \qquad (8.76)$$

where Ω_{sa} is the solid angle around which the spin is confined. Note Figure 8.2 is normalized by this factor: $\ln(\Omega) = \ln(4\pi)$.

Quantum case: In a similar fashion as before, considering the expansion of the hyperbolic sine function (Eq. (8.71)) and the Brillouin function for low values of x (Eq. (8.59)), the entropy is

$$s(x) \xrightarrow{x \to 0} \ln \left(\frac{ax + a^3 x^3/6}{bx + b^3 x^3/6} \right) - \frac{1}{3} \left(a^2 - b^2 \right) x^2 \tag{8.77}$$

and considering Eq.(8.73) we found

$$s(x) = \ln \left(\frac{a}{b} \right) + \ln \left(1 + \frac{1}{6} a^2 x^2 \right) - \ln \left(1 + \frac{1}{6} b^2 x^2 \right) - \frac{1}{3} \left(a^2 - b^2 \right) x^2, \tag{8.78}$$

$$= \ln \left(\frac{a}{b} \right) + \frac{1}{6} \left(a^2 - b^2 \right) x^2 - \frac{1}{3} \left(a^2 - b^2 \right) x^2. \tag{8.79}$$

Finally, the entropy reads as

$$s(x) = \ln \left(\frac{a}{b} \right) - \frac{x^2}{6} \left(a^2 - b^2 \right). \tag{8.80}$$

Higher-order correction is given by

$$+ \frac{x^4}{60} \left(a^4 - b^4 \right). \tag{8.81}$$

Here, this result also needs a careful analysis. Analogously to the classical case, the first term of the entropy is the logarithm of the number of accessible states, that, for this case, is

$$\Omega = \sum_{m_j = -j}^{+j} = 2j + 1 = \frac{a}{b} \tag{8.82}$$

and therefore the first term of this expansion is $\ln (2j + 1)$, as expected.

$\Longrightarrow$ *Limit* of high magnetic field and low temperature $(b \gg t \Rightarrow x \to \infty)$.

Classical case: To evaluate this limit we will need the previously derived limit of the Langevin function (Eq. (8.65))

$$s(x) \xrightarrow{x \to \infty} \ln (4\pi) + \ln \left(\frac{e^x - e^{-x}}{2} \right) - \ln (x) - x \left(1 - \frac{1}{x} \right) \tag{8.83}$$

and therefore

$$s(x) = \ln (2\pi) + 1 - \ln (x). \tag{8.84}$$

This limit is quite interesting. A simple inspection of the above equation leads us to

$$s(x) \xrightarrow{x \to \infty} -\infty \tag{8.85}$$

and therefore violates the third law of the thermodynamics! To remember, the third law says that the entropy must tend to zero when the temperature goes to zero. Here, the entropy goes to $-\infty$. Why? For this limit, high magnetic field and low temperature, the spins point along only one direction, i.e., along the magnetic field vector. Thus, how many occupied states do we have? For the classical case, not only one, but $d\Omega_{sa}$, i.e., an infinitesimal solid angle, that is smaller than the unit and therefore $\ln(d\Omega_{sa}) \to -\infty$. The correction of this failure is only possible to be fixed considering the quantum case, where the number of occupied states is really 1 and therefore $\ln(1) = 0$.

Quantum case: The limit of high x values must be handled carefully for the quantum case. As explained before (when evaluating the same limit for the magnetization), in addition to $x \to \infty$, we must consider $x \gg 2j$; otherwise, the argument of the hyperbolic cotangent function is not large enough.

Considering this approximation obtained for the Brillouin function (Eq. (8.67)), we have

$$s(x) \xrightarrow{x \to \infty} \ln\left(\frac{e^{+ax} - e^{-ax}}{e^{+bx} - e^{-bx}}\right) - x\left[1 + 2\left(ae^{-2ax} + be^{-2bx}\right)\right], \tag{8.86}$$

that is

$$s(x) = \ln\{\exp[(a-b)x]\} - x - 2x\left(ae^{-2ax} - be^{-2bx}\right) \tag{8.87}$$

and therefore, for $x \to \infty$, the entropy reads as

$$s(x) = 2x\left(be^{-2bx} - ae^{-2ax}\right) \xrightarrow{x \to \infty} 0. \tag{8.88}$$

These limits, for low and high values of x, are included in Figure 8.2.

8.1.2.4 Specific heat

Following the same idea of the previous quantities, the specific heat was already discussed for a general case. For a constant magnetic field it reads as (see Eq. (8.24))

$$c_b = x^2 \frac{\partial m(x)}{\partial x}. \tag{8.89}$$

Classical case: Considering then the first derivative of the Langevin function we found[3]

$$\frac{\partial}{\partial x} m(x) = -\operatorname{csch}^2(x) + \frac{1}{x^2} \tag{8.91}$$

and then the specific heat at constant magnetic field b reads as

$$c_b = 1 - x^2 \operatorname{csch}^2(x). \tag{8.92}$$

---

[3] $\dfrac{\partial}{\partial x} \coth(x) = -\operatorname{csch}^2(x) = -\dfrac{1}{\sinh^2(x)}.$ $\tag{8.90}$

The evaluation of c_m, the specific heat at constant magnetization, is left as an exercise to the reader.

Quantum case: Analogously to before, the first derivative of the Brillouin function is

$$\frac{\partial}{\partial x} m(x) = -a^2 \operatorname{csch}^2(ax) + b^2 \operatorname{csch}^2(bx) \tag{8.93}$$

and therefore

$$c_b = x^2 \left[b^2 \operatorname{csch}^2(bx) - a^2 \operatorname{csch}^2(ax) \right]. \tag{8.94}$$

It is important, before presenting the figures, to understand and predict what we expect for the behavior of the magnetic contribution to the specific heat. First, what is specific heat? Physically speaking, it is the amount of heat (energy) required to increase the temperature of a system in a certain interval. And how about the magnetic specific heat? It is therefore the amount of heat required by the magnetic moments to contribute to that task. Mathematically speaking, it is proportional to the first derivative of the entropy with respect to x times x (see Eq. (8.21)). As discussed above, for low values of x we expect that the entropy goes to a constant value ($\ln(\Omega)$), and indeed it occurs. Thus, the first derivative is zero, times x (that goes to zero), we predict that the magnetic specific heat, for $x \to 0$ limit, also must go to zero. On the other hand, for high values of x, the entropy also goes to a constant value, $\ln(1) = 0$, and therefore the first derivative is also zero; times $x \to \infty$ we have then an indetermination. It is simple to overcome this problem, because, from Eq. (8.88), we know that the entropy in this limit goes to zero exponentially (faster than x goes to infinity), and therefore c_b for high values of x must go to zero.

$\Longrightarrow$ *Limit* of low magnetic field and high temperature ($b \ll t \Rightarrow x \to 0$).

To evaluate this limit, we need the expansion of the quadratic hyperbolic cosecant function

$$\operatorname{csch}^2(x) \xrightarrow{x \to 0} -\frac{1}{3} + \frac{1}{x^2} + \frac{x^2}{15} + \mathcal{O}(x^4). \tag{8.95}$$

Classical case: Replacing the above expansion into Eq. (8.94), we have

$$c_b = 1 - x^2 \operatorname{csch}^2(x) \xrightarrow{x \to 0} 1 - x^2 \left(-\frac{1}{3} + \frac{1}{x^2} + \frac{x^2}{15} \right) \tag{8.96}$$

and then, it is possible to achieve this limit for the specific heat

$$c_b = \frac{x^2}{3} - \frac{x^4}{15}. \tag{8.97}$$

Note that c_b goes to zero when x also goes to zero, as expected.

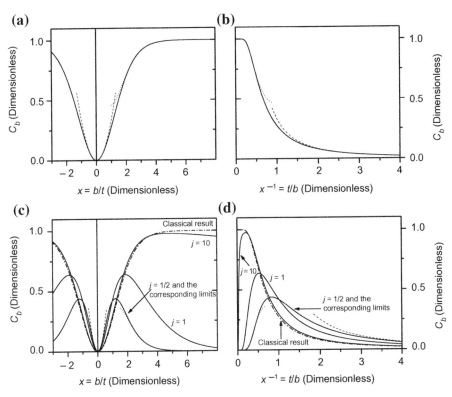

Figure 8.3 (a, c) Reduced specific heat at constant magnetic field c_b as a function of $x = b/t$. These figures can be seen as the specific heat measured at constant magnetic field, however, for several different values of b, and at constant temperature. (b, d) Reduced specific heat at constant magnetic field c_b as a function of $x^{-1} = t/b$. These figures can be seen as the influence of the temperature on c_b. Note (i) the limit for low magnetic field and high temperature ($x \to 0$ and $x^{-1} \to \infty$—dashed line), in which we only consider the first-order correction (in this case the term proportional to x^2) and (ii) the limit for high magnetic field and low temperature ($x \to \infty$ and $x^{-1} \to 0$—dotted line). See text for details. (a) Classical case. (b) Quantum case—the classical result (dashed-dotted line) is also presented, for the sake of comparison.

Quantum case: Considering also the expansion of $\operatorname{csch}^2(x)$, we then have

$$c_b(x) \xrightarrow{x \to 0} x^2 \left\{ b^2 \left[-\frac{1}{3} + \frac{1}{(bx)^2} + \frac{(bx)^2}{15} \right] - a^2 \left[-\frac{1}{3} + \frac{1}{(ax)^2} + \frac{(ax)^2}{15} \right] \right\}$$

(8.98)

and therefore, for low values of x

$$c_b(x) = \frac{x^2}{3} \left(a^2 - b^2 \right) - \frac{x^4}{15} \left(a^4 - b^4 \right).$$

(8.99)

It is also important to see that c_b goes to zero when x goes to zero, as expected.

$\Longrightarrow$ *Limit* of high magnetic field and low temperature ($b \gg t \Rightarrow x \to \infty$). This limit is of easy evaluation. Considering

$$\text{csch}^2(x) = \frac{1}{\sinh^2(x)} = \frac{4}{\left(e^x - e^{-x}\right)^2} \xrightarrow{x \to \infty} 4e^{-2x}, \tag{8.100}$$

it is possible to obtain c_b for both cases.

Classical case: From the above result, we have for high values of x

$$c_b = 1 - x^2 4e^{-2x}. \tag{8.101}$$

Note for this case, when x goes to infinity, c_b goes to 1, instead of the expected zero. It is also a consequence of the failure of the third law of the thermodynamics (see entropy discussion above). This problem is overcome by evaluating the quantum case.

Quantum case: Again, considering the expansion on Eq. (8.100)

$$c_b(x) = 4x^2 \left[b^2 e^{-2bx} - a^2 e^{-2ax} \right]. \tag{8.102}$$

In this case, as expected, c_b goes to zero when x tends to high values. Evaluation of c_m is left as an exercise to the reader.

See Figure 8.3 for a better understanding of the text.

8.2   Itinerant paramagnetism

To evaluate the paramagnetic properties of an electrons gas, we only need to consider the kinetic energy (already discussed in Chapter 6) and magnetic moments of those particles. It is therefore a step forward in comparison with Chapter 7, where the dia-magnetism was discussed, based only on the kinetics of those electrons. Before going further on this chapter, we strongly recommend to read Chapter 6, where we discussed a free electron gas, without magnetic moment and without an applied magnetic field.

As already discussed in Chapter 6, each $|k\rangle$ state can accommodate two electrons: one with magnetic moment up ($\Uparrow$) and other with magnetic moment down ($\Downarrow$). Thus, let us divide this electrons gas in two subbands.

When an external magnetic field of induction $\vec{B} = B\hat{k}$ is applied, both subbands change energy; in $-\mu_B B$ for the "up" subband and $+\mu_B B$ for the "down" subband, as discussed in the exercises of Chapter 3. Thus

$$E_{\Uparrow} = E - \mu_B B \quad \text{and} \quad E_{\Downarrow} = E + \mu_B B. \tag{8.103}$$

To evaluate any thermodynamic property, we must know the *Grand* canonical potential $\Phi = -\frac{1}{\beta} q$ and therefore we need to compute $q = \ln \mathcal{Z}$ (where $\mathcal{Z}$ is the *Grand* partition function). However, from Chapter 6, we know that

$$q = \int_0^\infty g(E) \ln \left(1 + z e^{-\beta E} \right) dE$$

and therefore it is possible to evaluate q for each subband after a simple change of variables. First, we must include in the above equation the information that the density of states is shifted; and thus for the "up" subband it reads as

$$q_\uparrow(z_\uparrow) = \int_{-\mu_B B}^{\infty} g(E_\uparrow + \mu_B B) \ln\left(1 + z_\uparrow e^{-\beta E_\uparrow}\right) dE_\uparrow. \tag{8.104}$$

Now, we must change the variables. Considering $E' = E_\uparrow + \mu_B B$ the above equation can be rewritten as

$$q_\uparrow(z) = \int_0^{\infty} g(E') \ln\left(1 + z' e^{-\beta E'}\right) dE', \tag{8.105}$$

where

$$z' = z_\uparrow e^{\beta \mu_B B} = \exp[\beta(\mu_\uparrow + \mu_B B)] = e^{\beta \mu} \tag{8.106}$$

and μ above written means the chemical potential of an electrons gas without magnetic field. Thus, it is possible to write

$$\mu(N_\uparrow) = \mu_\uparrow(N_\uparrow) + \mu_B B \tag{8.107}$$

The above result is quite important, because we have already developed the properties of an electrons gas without magnetic field (see Chapter 6).

Analogously,

$$q_\downarrow(z_\downarrow) = \int_{+\mu_B B}^{\infty} g(E_\downarrow - \mu_B B) \ln\left(1 + z_\downarrow e^{-\beta E_\downarrow}\right) dE_\downarrow \tag{8.108}$$

and also changing the variables as $E'' = E_\downarrow - \mu_B B$ we obtain

$$q_\downarrow(z) = \int_0^{\infty} g(E'') \ln\left(1 + z'' e^{-\beta E''}\right) dE'', \tag{8.109}$$

where

$$z'' = z_\downarrow e^{-\beta \mu_B B} = exp[\beta(\mu_\downarrow - \mu_B B)] = e^{\beta \mu} \tag{8.110}$$

and then

$$\mu(N_\downarrow) = \mu_\downarrow(N_\downarrow) - \mu_B B. \tag{8.111}$$

However, the number of particles N in the system is constant. The amount of particles that leave the "down" subband to go into the "up" subband, when a magnetic field is applied, is the same and therefore it is possible to consider $\mu_\uparrow(N_\uparrow) = \mu_\downarrow(N_\downarrow)$. Thus, considering Eqs. (8.107) and (8.111), it is possible to write

$$\mu(N_\uparrow) - \mu(N_\downarrow) = 2\mu_B B \,. \tag{8.112}$$

The above equation is of great importance, since from that we will obtain the magnetization and magnetic susceptibility of this electrons gas. Now, it is quite easy to proceed, since we already know, from Chapter 6, the chemical potential μ of an electrons gas at low temperature ($\epsilon_F \gg k_B T$) and high temperature ($\epsilon_F \ll k_B T$) limits. However, before we going further, let us write the magnetization of the system as a function of the number of electrons in the "up" and "down" subbands, i.e., $N_\Uparrow$ and $N_\Downarrow$, respectively.

The magnetization can be written as

$$M = M_\Uparrow + M_\Downarrow,$$ (8.113)

but it is quite easy to show that it is also possible to write the magnetization as

$$M = (N_\Uparrow - N_\Downarrow)\mu_B$$ (8.114)

and this proof is left as an exercise to the reader. On the other hand, the total number of particles N must be constant, i.e., $N = N_\Uparrow + N_\Downarrow$ and therefore the following relations hold:

$$\frac{2N_\Uparrow}{N} = (1 + m) \quad \text{and} \quad \frac{2N_\Downarrow}{N} = (1 - m),$$ (8.115)

where $m = M/N\mu_B$ is a reduced and dimensionless magnetization.

From now on, let us develop the low and high temperature properties of this paramagnetic electrons gas.

$\Longrightarrow$ *Limit* of low temperature ($\epsilon_F \gg k_B T$).

This limit of low temperature was already developed in Chapter 6 and therefore the chemical potential is (from Eq. (6.39))

$$\mu(N) \approx \epsilon_F \left[1 - \frac{\pi^2}{12} \left(\frac{k_B T}{\epsilon_F} \right)^2 \right],$$

where (from Eq. (6.30))

$$\epsilon_F = \left(\frac{N}{V} \frac{6\pi^2}{\bar{g}} \right)^{2/3} \frac{\hbar^2}{2m}.$$

In this sense, $\mu(N_\Uparrow)$ is easily obtained just replacing N by $N_\Uparrow$ and considering $\bar{g} = 1$, since all of the magnetic moments of this subband have only one projection (up). Thus

$$\mu(N_\Uparrow) \approx \epsilon_{F\Uparrow} \left[1 - \frac{\pi^2}{12} \left(\frac{k_B T}{\epsilon_{F\Uparrow}} \right)^2 \right],$$ (8.116)

where

$$\epsilon_{F\Uparrow} = \left(\frac{N_\Uparrow}{V}6\pi^2\right)^{2/3}\frac{\hbar^2}{2m}$$

$$= \left(\frac{2N_\Uparrow}{N}\right)^{2/3}\epsilon_{F0}$$

$$= (1+m)^{2/3}\epsilon_{F0}. \tag{8.117}$$

Above,

$$\epsilon_{F0} = \left(\frac{N}{V}3\pi^2\right)^{2/3}\frac{\hbar^2}{2m} \tag{8.118}$$

stands for the zero field Fermi energy, where we considered $\bar{g} = 1$ and $N \to N/2$ (at zero field $N_\Uparrow = N_\Downarrow = N/2$). Also note that ϵ_{F0} is equal to the Fermi energy considering only one band, i.e., $\epsilon_F(N, \bar{g} = 2)$. This fact is reasonable. Let us suppose that, filling the energy levels, we only include electrons with spins up, from the bottom level up to the highest one (thus, ϵ_{F0}). After including $N/2$ electrons, we start to include the other half, with spin down electrons. In this case, it is possible to restart to fill from the bottom up to ϵ_{F0}, and fill with the remaining $N/2$ electrons (completing therefore the N electrons). Thus, $\epsilon_{F0} = \epsilon_F(N, \bar{g} = 2) = \epsilon_F(N/2, \bar{g} = 1)$.

Analogously to the "up" subband calculi, the same idea can be used to the "down" subband

$$\mu(N_\Downarrow) \approx \epsilon_{F\Downarrow}\left[1 - \frac{\pi^2}{12}\left(\frac{k_B T}{\epsilon_{F\Downarrow}}\right)^2\right], \tag{8.119}$$

where

$$\epsilon_{F\Downarrow} = (1-m)^{2/3}\epsilon_{F0} \tag{8.120}$$

Thus, Eq. (8.112) can be rewritten as

$$\epsilon_{F\Uparrow}\left[1 - \frac{\pi^2}{12}\left(\frac{k_B T}{\epsilon_{F\Uparrow}}\right)^2\right] - \epsilon_{F\Downarrow}\left[1 - \frac{\pi^2}{12}\left(\frac{k_B T}{\epsilon_{F\Uparrow}}\right)^2\right] = 2\mu_B B. \tag{8.121}$$

After some changes the above equation reads as

$$(1+m)^{2/3} - (1-m)^{2/3} - \frac{\pi^2}{12}t^2\left[(1+m)^{-2/3} - (1-m)^{-2/3}\right] = 2b, \tag{8.122}$$

where $b = \mu_B B/\epsilon_{F0}$ and $t = k_B T/\epsilon_{F0}$ are the reduced and dimensionless magnetic field and temperature, respectively. The above equation is the "equation of state" for this system, for the low temperature limit ($t \to 0$). Note that this result was achieved considering the properties of an electrons gas, already discussed in Chapter 6.

For $t = 0$ (and even $t \to 0$), it is easy to see that for $b = 0$ the unique possible solution is $m = 0$. Thus, without external magnetic field there is no magnetization, like in the diamagnetic case. For $b \to 0$ limit, the magnetization is therefore quite small ($m \to 0$) and the equation of state can be rewritten as[4]

$$m = \frac{3}{2}b\left[1 + \frac{\pi^2}{12}t^2\right]^{-1} \tag{8.123}$$

and therefore

$$m = \frac{3}{2}b\left[1 - \frac{\pi^2}{12}t^2\right]. \tag{8.124}$$

The reduced and dimensionless magnetic susceptibility $s = \chi\epsilon_{F0}/(\mu_0 N\mu_B^2)$ can also be easily obtained and holds as

$$s = \frac{3}{2}\left[1 - \frac{\pi^2}{12}t^2\right]. \tag{8.125}$$

In terms of dimensional variables, instead of those dimensionless, the paramagnetic susceptibility at the low temperature regime reads as

$$\chi_p = N\left(\frac{3}{2}\frac{\mu_B}{\epsilon_{F0}}\right)\mu_0\mu_B\left[1 - \frac{\pi^2}{12}\left(\frac{k_B T}{\epsilon_{F0}}\right)^2\right]. \tag{8.126}$$

However, for absolute zero temperature ($T = 0$) and even $k_B T \ll \epsilon_{F0}$, the magnetic susceptibility is

$$\chi_p = N\left(\frac{3}{2}\frac{\mu_B}{\epsilon_{F0}}\right)\mu_0\mu_B = -3\chi_d, \tag{8.127}$$

where χ_d is the Landau diamagnetic susceptibility, given by Eq. (7.41). Note the above result is temperature independent. In terms of the density of states (Eq. (6.32)), the above equation reads as

$$\chi_p = [g(\epsilon_{F0})\mu_B]\mu_0\mu_B. \tag{8.128}$$

When the magnetic susceptibility is written as above, it depends directly on the density of state at the Fermi level and this fact is of great importance when we study paramagnetic (and ferromagnetic) materials. The above result is the well-known Pauli paramagnetic susceptibility.

Another interesting point is the following: this result comes from the magnetic properties of an electrons gas with magnetic moment; while the itinerant diamagnetic result (Eq. (7.41)) comes from the same electrons gas without magnetic moment and

[4]For the $x \to 0$ limit, the following approximation is valid: $(1 + bx)^n = 1 + nbx + \cdots$

considering only the Landau levels (due to the magnetic field). It is possible to see that $\chi_d = -\frac{1}{3}\chi_p$, and the total magnetic susceptibility (considering both, electronic angular momenta and the Landau levels) is

$$\chi_T = \chi_d + \chi_p = \frac{2}{3}\chi_p > 0. \tag{8.129}$$

$\Longrightarrow$ *Limit* of high temperature ($\epsilon_F \ll k_B T$)

In the same fashion as before, we will take advantage of the results already discussed in Chapter 6, where the chemical potential μ of an electrons gas was already developed under several conditions. Thus, at high temperature ($\epsilon_F \ll k_B T$), it reads as

$$\mu(N) = \beta^{-1} \ln \left[\frac{N}{V} \frac{1}{\bar{g}} \left(\frac{4\pi \hbar^2 \beta}{2m} \right)^{3/2} \right] \tag{8.130}$$

and, since we need $\mu(N_\uparrow)$ instead of $\mu(N)$, we only need to substitute N by $N_\uparrow$ and then

$$\mu(N_\uparrow) = \beta^{-1} \ln \left[\frac{2N_\Uparrow}{N} \xi \right], \tag{8.131}$$

where

$$\xi = \frac{1}{2} \frac{N}{V} \left(\frac{4\pi \hbar^2 \beta}{2m} \right)^{3/2}. \tag{8.132}$$

Note we considered $\bar{g} = 1$, since we are dealing with only the "up" subband and the magnetic moments have only one projection.

Analogously,

$$\mu(N_\downarrow) = \beta^{-1} \ln \left[\frac{2N_\Downarrow}{N} \xi \right] \tag{8.133}$$

and therefore Eq. (8.112) can be rewritten as

$$\beta^{-1} \ln \left[\frac{2N_\Uparrow}{N} \xi \right] - \beta^{-1} \ln \left[\frac{2N_\Downarrow}{N} \xi \right] = 2\mu_B B. \tag{8.134}$$

Simple calculations resume the above equation as

$$\ln \left[\frac{1+m}{1-m} \right] = 2\frac{b}{t} \tag{8.135}$$

and therefore it is possible to find the final equation of state for this limit of high temperature

$$m = \tanh \left(\frac{b}{t} \right). \tag{8.136}$$

Analogously to the low temperature case, it is easy to see that for $b = 0$ there is no magnetization. On the other hand, for the low field approximation, it is possible to write[5]

$$m = \frac{b}{t}$$

(8.137)

and then the reduced magnetic susceptibility is

$$s = \frac{1}{t}.$$

(8.138)

Finally, in terms of dimensional quantities, the paramagnetic susceptibility at the high temperature regime is

$$\chi_p = N \left(\frac{\mu_B}{k_B T} \right) \mu_0 \mu_B = -3\chi_d,$$

(8.139)

where the above diamagnetic susceptibility is given by Eq. (7.48). Note that at high temperature the relation between χ_p and χ_d remains the same, i.e., $\chi_p = -3\chi_d$.

Figures 8.4 and 8.5 present, respectively, the magnetization and susceptibility of a paramagnetic electrons gas under low and high temperature regimes.

8.3 van Vleck paramagnetism

The results of this section are of great importance to those cases where we cannot achieve an analytical solution to the eigenvalues of energy. The idea is to use perturbation theory to obtain the first and second corrections on the energy spectra and then write the magnetic susceptibility.

Let us start from the Zeeman energy of the ith state

$$\varepsilon_i = -\vec{\mu}_i \cdot \vec{B},$$

(8.140)

that can be rewritten as

$$\varepsilon_i = -\sum_u (\mu_i)_u B_u,$$

(8.141)

where $u = x, y, z$. Then, the magnetic moment is

$$(\mu_i)_u = -\frac{\partial \varepsilon_i}{\partial B_u}.$$

(8.142)

To simplify the problem, let us consider $\vec{B} = B_u \hat{u}$, i.e., $\vec{B}$ is along u (that is either x, y, or z). Thus

$$\varepsilon_i = -(\mu_i)_u B_u.$$

(8.143)

---

[5]For the $x \to 0$ limit, the following approximation is valid: $\tanh(x) = x - x^3/3 + \cdots$

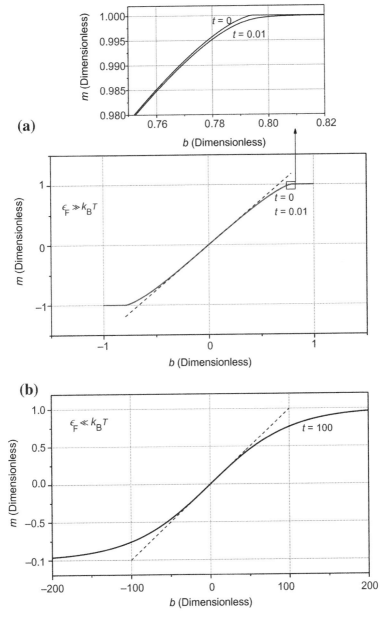

Figure 8.4 Paramagnetic contribution to the magnetization. Each panel above corresponds to a certain approximation. These are displayed via the reduced and dimensionless magnetization $m = M/N\mu_B$, magnetic field $b = \mu_B B/\epsilon_{F0}$, and temperature $t = k_B T/\epsilon_{F0}$, where $\epsilon_{F0} = \epsilon_F(\bar{g} = 2)$ is the zero field Fermi energy. (a) Solid line: low temperature ($\epsilon_F \gg k_B T$) approximation (see Eq. (8.122) for details). Dashed line: low magnetic field approximation and $t = 0$ (see Eq. (8.124) for details). (b) Solid line: high temperature ($\epsilon_F \ll k_B T$) approximation (see Eq. (8.136) for details). Dashed line: low magnetic field approximation and $t = 0$ (see Eq. (8.137) for details).

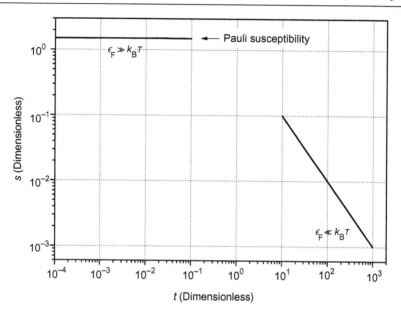

Figure 8.5 Reduced paramagnetic susceptibility $s = \chi \epsilon_{F0}/(\mu_0 N \mu_B^2)$ for low ($\epsilon_F \gg k_B T$) and high ($\epsilon_F \ll k_B T$) temperature limits. See Eqs. (8.125) and (8.138) for details.

We know from Eqs. (5.29) and (5.40), that

$$\frac{M_u}{N} = \mu_u = \frac{\sum_i (\mu_i)_u e^{-\varepsilon_i \beta}}{\sum_i e^{-\varepsilon_i \beta}} \tag{8.144}$$

and then:

$$\frac{M_u}{N} = \frac{\sum_i \left(-\partial \varepsilon_i / \partial B_u\right) e^{-\varepsilon_i \beta}}{\sum_i e^{-\varepsilon_i \beta}}. \tag{8.145}$$

The idea of van Vleck was to consider the energy as a function of powers of the magnetic field B_u

$$\varepsilon_i = \varepsilon_i^{(0)} + \varepsilon_i^{(1)} B_u + \varepsilon_i^{(2)} B_u^2 + \cdots \tag{8.146}$$

From the above, the magnetic moment holds as

$$(\mu_i)_u = -\varepsilon_i^{(1)} - 2\varepsilon_i^{(2)} B_u - \cdots \tag{8.147}$$

Other approximation must be considered: the magnetic energy is smaller than the thermal energy. Thus,

$$e^{-\varepsilon_i \beta} = \exp\left[-\left(\varepsilon_i^{(0)} + \varepsilon_i^{(1)} B_u\right)\beta\right] = e^{-\varepsilon_i^{(0)} \beta}\left[1 - \varepsilon_i^{(1)} B_u \beta\right]. \tag{8.148}$$

Placing these two approximations into Eq. (8.144), then

$$\frac{M_u}{N} = \mu_u = \frac{\sum_i \left(-\varepsilon_i^{(1)} - 2\varepsilon_i^{(2)} B_u\right) e^{-\varepsilon_i^{(0)}\beta} \left[1 - \varepsilon_i^{(1)} B_u \beta\right]}{\sum_i e^{-\varepsilon_i^{(0)}\beta} \left[1 - \varepsilon_i^{(1)} B_u \beta\right]}. \tag{8.149}$$

The above result can be further simplified: (i) do not consider terms on B_u^2 and (ii) the magnetization at zero field is zero, i.e.,

$$M_u(B_u = 0) \propto \sum_i \varepsilon_i^{(1)} e^{-\varepsilon_i^{(0)}\beta} = 0. \tag{8.150}$$

Thus:

$$\frac{M_u}{N} = B_u \frac{\sum_i \left[\left(\varepsilon_i^{(1)}\right)^2 \beta - 2\varepsilon_i^{(2)}\right] e^{-\varepsilon_i^{(0)}\beta}}{\sum_i e^{-\varepsilon_i^{(0)}\beta}}. \tag{8.151}$$

Now, the susceptibility is defined in accordance with Eq. (5.42); however, for the specific case $v = u$

$$\chi_{uu} = \lim_{B_u \to 0} \mu_0 \frac{\partial M_u}{\partial B_u} \tag{8.152}$$

and then we obtain the famous van Vleck susceptibility

$$\chi_{uu} = N\mu_0 \frac{\sum_i \left[\left(\varepsilon_i^{(1)}\right)^2 \beta - 2\varepsilon_i^{(2)}\right] e^{-\varepsilon_i^{(0)}\beta}}{\sum_i e^{-\varepsilon_i^{(0)}\beta}}. \tag{8.153}$$

The energy corrections $\varepsilon_i^{(1)}$ and $\varepsilon_i^{(2)}$ are obtained from perturbation theory (see Complement 8.A). Further discussion of the consequences of the van Vleck susceptibility will be given in Complement 8.B.

Complements

8.A   Perturbation theory

First, let us consider a Hamiltonian $\mathcal{H}_0$, diagonal on the basis $|n^{(0)}\rangle$. Thus, it is possible to write

$$\mathcal{H}_0 |n^{(0)}\rangle = \varepsilon_n^{(0)} |n^{(0)}\rangle, \tag{8.A.1}$$

where $\varepsilon_n^{(0)}$ are the eigenvalues of this Hamiltonian. Now, let us introduce a perturbation $\mathcal{W}$, where the intensity of this perturbation is ruled by a parameter $0 \le \lambda \le 1$. Thus, the total Hamiltonian can be written as

$$\mathcal{H} = \mathcal{H}_0 + \lambda \mathcal{W}. \tag{8.A.2}$$

An important observation: $\mathcal{W}$ is also written on the basis $|n^{(0)}\rangle$ and, for the results shown below, it must be Hermitian. Since $\mathcal{W}$ is not diagonal, the total Hamiltonian is also nondiagonal (on the $|n^{(0)}\rangle$ basis). To overcome this problem, it is possible to say that $\mathcal{H}$ is diagonal on the basis $|n\rangle$ and therefore, we can write

$$\mathcal{H}|n\rangle = \varepsilon_n |n\rangle, \tag{8.A.3}$$

where ε_n are the eigenvalues of the total Hamiltonian.

Now, what we need to know are the new eigenvalues of energy ε_n and the new eigenvectors $|n\rangle$. To this purpose, perturbation theory considers both as an expansion on the parameter λ. Thus,

$$|n\rangle = |n^{(0)}\rangle + \lambda |n^{(1)}\rangle + \lambda^2 |n^{(2)}\rangle + \cdots \tag{8.A.4}$$

and

$$\varepsilon_n = \varepsilon_n^{(0)} + \lambda \varepsilon_n^{(1)} + \lambda^2 \varepsilon_n^{(2)} + \cdots \tag{8.A.5}$$

8.A.1   Nondegenerated case

Let us then consider the nondegenerated case of the nonperturbed Hamiltonian, i.e., for each eigenvalue of energy $\varepsilon_n^{(0)}$ there is only one eigenvector $|n^{(0)}\rangle$. After some steps (not shown), it is possible to obtain the first- and second-order correction to the energy

$$\varepsilon_n^{(1)} = \langle n^{(0)}|\mathcal{W}|n^{(0)}\rangle \tag{8.A.6}$$

and

$$\varepsilon_n^{(2)} = \sum_{k \neq n} \frac{\left|\langle k^{(0)}|\mathcal{W}|n^{(0)}\rangle\right|^2}{\varepsilon_n^{(0)} - \varepsilon_k^{(0)}}. \tag{8.A.7}$$

Note that the zero-order correction ($\varepsilon_n^{(0)}$) is previously known, since these are the eigenvalues of the nonperturbed Hamiltonian $\mathcal{H}_0$. The first correction $\varepsilon_n^{(1)}$ corresponds to the diagonal values of $\mathcal{W}$ on the basis $|n^{(0)}\rangle$ (i.e., as the perturbation was written originally). In what concerns the second-order correction $\varepsilon_n^{(2)}$, this contribution depends on the off-diagonal elements of the perturbation and, physically speaking, represents a mixture of the n states of the system.

The first-order correction to the eigenvectors is

$$|n^{(1)}\rangle = \sum_{k \neq n} \frac{\langle k^{(0)}|\mathcal{W}|n^{(0)}\rangle}{\varepsilon_n^{(0)} - \varepsilon_k^{(0)}} |k^{(0)}\rangle. \tag{8.A.8}$$

8.A.2   Degenerated case

This case is quite different from the previous one. Now, there is more than one eigenvector $|n^{(0)}\rangle$ with the same eigenvalue of energy. From this fact immediately arises a problem, since it creates a zero denominator for Eq. (8.A.7), i.e., an infinity second-order correction of energy (physically impossible). Thus, there is a need of a small change on the theory (not shown) and now the corrections for the energy are

$$\varepsilon_n^{(1)} = \langle l^{(0)}|\mathcal{W}|l^{(0)}\rangle \tag{8.A.9}$$

and

$$\varepsilon_n^{(2)} = \sum_{\substack{k \notin D \\ n \in D}} \frac{\left|\langle k^{(0)}|\mathcal{W}|n^{(0)}\rangle\right|^2}{\varepsilon_n^{(0)} - \varepsilon_k^{(0)}}. \tag{8.A.10}$$

Note the similarity among the results above presented and those for the nondegenerated case. The difference is quite small, but takes care of the divergence of the second-order correction to the energy.

For this case, the zeroth-order term $\varepsilon_n^{(0)}$ remains the same for the nondegenerated case, i.e., the eigenvalues of the diagonal and nonperturbed Hamiltonian $\mathcal{H}_0$. The first-order correction is not obtained from the diagonal of the perturbation matrix written on the basis $|n^{(0)}\rangle$, but in a subspace D in which are the degenerated states. The procedure, for this case, is to isolate a submatrix of $\mathcal{H}_0$ that contains the degenerated eigenvalues and then map it on the $\mathcal{W}$ matrix. After isolating this submatrix on the perturbed matrix, the procedure is to diagonalize it, obtaining thus the first-order correction of energy for those levels previously degenerated. Generally, this first-order correction is enough to rise the degeneracy. In what concerns the second-order correction, that has the problem of the singularity (zero denominator), there is also a difference on the evaluation, comparing to the nondegenerated case. Instead of the index k (corresponding to the matrix lines) runs over all of the elements, it runs only for those out of the subspace D.

8.B   Rare-earth paramagnetism: an example

As discussed in Chapter 3, for the rare-earths the most energetic electrons lie in the $4f$ shell; however, it is an inner shell and therefore shielded from the environment (see Figure 3.6). Thus, the magnetic properties of these ions came from the well-defined angular momentum $\vec{L}$, spin momentum $\vec{S}$, and total angular momentum $\vec{J}$; where these are determined by Hund's rules, as discussed in Chapter 3. It is recommended to review that chapter before going further in this section.

From the third Hund's rule, the energy of each j multiplet is given, in a first approximation, by the spin–orbit coupling

$$\mathcal{H}_{SL} = \zeta \vec{S} \cdot \vec{L}. \tag{8.B.11}$$

However, the total angular momentum is

$$\vec{J} = \vec{L} + \vec{S} \tag{8.B.12}$$

and therefore

$$J^2 = L^2 + S^2 + 2\vec{S} \cdot \vec{L}. \tag{8.B.13}$$

From the above equation, the spin–orbit Hamiltonian can be rewritten as

$$\mathcal{H}_{SL} = \frac{\zeta}{2}\left(J^2 - L^2 - S^2\right) \tag{8.B.14}$$

and the eigenvalues of energy assume

$$\varepsilon_j = \frac{\zeta}{2}\left[j(j+1) - l(l+1) - s(s+1)\right]. \tag{8.B.15}$$

Since both, l and s are constants for each ion, we can shift these energy spectra to

$$\varepsilon_j = \frac{\zeta}{2}j(j+1). \tag{8.B.16}$$

As discussed in Chapter 3, the values of j range from $j = |s-l|$ to $j = s+l$. Remember, $j = |s-l|$ corresponds to the ground state for light rare-earths, i.e., those with $4f$ shell less than half-filled (from Cerium-Ce up to Europium-Eu), while $j = s+l$ is the ground state for heavy rare-earths, i.e., those with $4f$ shell more than half-filled (from Terbium-Tb up to Ytterbium-Yb).

Thus, from Eq. (8.B.16), it is possible to see, at least in first approximation, that the energy of each multiplet is proportional to the spin–orbit parameter ζ. However, each ion has its own ζ and $\{j\}$ set and therefore the splitting between the multiplets is different for each ion (as expected). The most important point for the present discussion is the energy separation Δ between the ground state and the first excited state—these values are presented in Table 3.2 and, for a better understanding, in Figure 8.6 bottom, where the energy levels can be compared to the room temperature (dashed line—300 K). Note therefore that we can divide the rare-earth group in two: those with spin–orbit parameter ζ comparable to the room temperature energy and those with spin–orbit parameter much bigger than the room temperature. For the first group (mainly trivalent Europium and Samarium), it is needed to consider the first excited state (and even the other states, for some cases), when evaluating thermodynamic quantities; while for the other rare-earths, only the ground state ($j = |s-l|$ for light rare-earths and $j = s+l$ for heavy rare-earths) is enough to describe thermodynamic quantities.

Let us consider the magnetic susceptibility at room temperature of these trivalent ions, as presented in Figure 8.6-top. This figure considers some cases: only the ground state to the susceptibility ("Hund" label), the excited states to the susceptibility ("van Vleck" label), and the experimental susceptibility ("experimental" label). Note "Hund" case does not match the "experimental" case for those ions in which the room temperature is comparable to the spin–orbit parameter ζ (Europium and Samarium). On the contrary, for these ions, the "van Vleck" cases approach the experimental result, and this result ratifies the needed to consider the excited states when evaluating the thermodynamic quantities of Europium and Samarium. Next subsections are devoted to the quantitative description of the magnetic susceptibility for both cases: $\zeta \approx k_B T$ and $\zeta \gg k_B T$.

8.B.1 Low values of spin–orbit parameter ($\approx k_B T$)

To consider the excited states to the magnetic susceptibility, the van Vleck susceptibility (Eq. (8.153)), is the best starting point. Considering the Zeeman term as a perturbation,

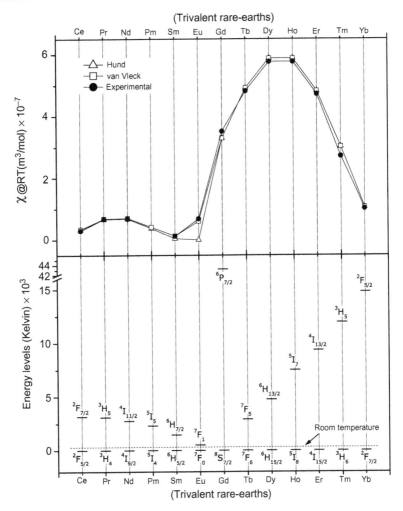

Figure 8.6 Ground state and first excited state (bottom) and magnetic susceptibility at room temperature (top) for trivalent rare-earths. See text for details on the meaning of the labels.

it is possible to find (for a deeper development, see references [2–4])

$$\chi = N\mu_0\mu_B^2 \frac{\sum_j (2j+1)\chi_j e^{-\varepsilon_j/k_B T}}{\sum_j (2j+1)e^{-\varepsilon_j/k_B T}}, \tag{8.B.17}$$

where j runs from $|s - l|$ up to $s + l$,

$$\chi_j = \frac{g_j^2 j(j+1)}{3k_B T} + \alpha_j \tag{8.B.18}$$

and g_j is given by Eq. (3.51). In addition,

$$\alpha_j = \frac{1}{6\zeta(2j+1)}\left[\frac{F(j+1)}{j+1} - \frac{F(j)}{j}\right] \tag{8.B.19}$$

and

$$F(j) = \frac{1}{j}\left[(l+s+1)^2 - j^2\right]\left[j^2 - (s-l)^2\right]. \tag{8.B.20}$$

As an example, for the trivalent Europium case, the magnetic susceptibility considering all of the excited states reads as

$$\chi(R^{3+}) = N\mu_0 \frac{\mu_B^2}{3\zeta} \frac{\mathcal{N}}{\mathcal{D}}, \tag{8.B.21}$$

where, for the case of Europium

$$\mathcal{N} = 24 + \left(\frac{27}{2}x - \frac{3}{2}\right)e^{-x} + \left(\frac{135}{2}x - \frac{5}{2}\right)e^{-3x} + \left(189x - \frac{7}{2}\right)e^{-6x}$$
$$+ \left(405x - \frac{9}{2}\right)e^{-10x} + \left(\frac{1485}{2}x - \frac{11}{2}\right)e^{-15x} + \left(\frac{2457}{2}x - \frac{13}{2}\right)e^{-21x}, \tag{8.B.22}$$

$$\mathcal{D} = 1 + 3e^{-x} + 5e^{-3x} + 7e^{-6x} + 9e^{-10x} + 11e^{-15x} + 13e^{-21x}, \tag{8.B.23}$$

and $x = \zeta/k_B T$.

Now, there is an interesting point that needs further attention. Note when $l = s$ there is singlet ($j = 0$) and therefore an indetermination evaluating g_j (Eq. (3.51)) and α_j (Eq. (8.B.19)). How to overcome this point is left as an exercise to the reader.

Figure 8.7a presents the experimental magnetic susceptibility of a trivalent europium compound (EuBO$_3$), as well as the van Vleck susceptibility (Eq. (8.B.17), considering only the ground state ($^7F_0$), the ground state, and the first excited state ($^7F_{\{0,1\}}$), and all of the available states ($^7F_{\{0,6\}}$). Note therefore the first excited state (at least) is needed to describe the magnetic susceptibility of this compound. The theoretical curve fails in a limited range of temperatures due to other ingredients not introduced in the model, like magnetocrystalline anisotropy.

8.B.2   High values of spin–orbit parameter ($\zeta \gg k_B T$)

This case is the simplest one. Since the spin–orbit parameter is bigger than the room temperature, it is possible to consider only the ground state, i.e., a unique value of j. This assumption is valid for all of the rare-earths, except for those mentioned before: Europium and Samarium (see Figure 8.6).

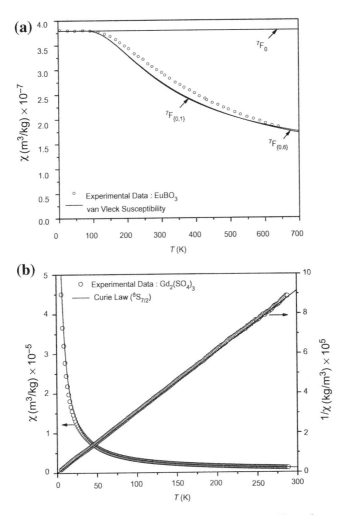

Figure 8.7 Experimental and theoretical paramagnetic susceptibility of some rare-earth compounds. (a) Low value of spin–orbit parameter example. Experimental data (open circles) of a paramagnetic trivalent europium compound (EuBO$_3$) and the van Vleck susceptibility (solid lines – see equation 8.B.17), considering only the ground state ($^7F_0$), the ground state and the first excited state ($^7F_{0,1}$) and all of the available states ($^7F_{0,6}$). The spin–orbit parameter considered was $\zeta = 471$K. (b) High value of spin–orbit parameter example. Experimental data (open circles) of a paramagnetic trivalent gadolinium compound (Gd$_2$(SO$_4$)$_3$) and the Curie law (solid line—see equation 8.61), considering only the $^8S_{7=2}$ ground state. The left axis represents the inverse magnetic susceptibility, from which it is possible to obtain the paramagnetic effective moment (equation 8.63), found to be 7.91(7) μ_B/Gd.

Thus, the thermodynamic quantities for this case are those described in Section 8.1. As an example, see Figure 8.7b, in which the experimental magnetic susceptibility of a trivalent gadolinium compound ($Gd_2(SO_4)_3$) is presented. It is quite well described by the Curie law (see Eq. (8.61)), considering only the $^8S_{7/2}$ ground state. The inverse magnetic susceptibility provides information on the paramagnetic effective moment (Eq. (8.63)), found to be 7.91(7) μ_B/Gd. This value agrees with the predicted one: 7.94 μ_B/Gd, considering $l = 0$, $s = j = 7/2$, and $g = 2$.

Part Three
Cooperative Magnetism

9 Magnetic Interactions

The previous chapters dealt with noninteracting magnetic moments and therefore non-cooperative orderings. To obtain spontaneous magnetization, due to the cooperative ordering, magnetic moments need to interact among them. Thus, this is the aim of this chapter, to introduce the most important ways of interactions.

9.1 Direct exchange

This section discusses the direct interaction between magnetic moments, starting from a simple model: two electrons and two closed shell atoms.

9.1.1 A simple model

First, let us consider one closed shell atom and one extra electron. We know that the Hamiltonian that describes this system can be written as

$$\mathcal{H}_{11} = -\frac{\hbar^2}{2m}\nabla^2 + V_0(r), \tag{9.1}$$

where the first term represents the kinetic energy of the electron and the second term is the potential due to the atom. Following this idea, we can add one extra closed shell atom, close to the first; however, still considering only one electron. Thus, the Hamiltonian reads as

$$\mathcal{H}_{21} = \mathcal{H}_{11}(r) + V_0(|r - R|). \tag{9.2}$$

Note the second term is the potential due to the added atom. Now, let us add an extra electron and then the total Hamiltonian is

$$\mathcal{H}_{22} = \mathcal{H}_{21}(r_1) + \mathcal{H}_{21}(r_2) + V_c(r_1, r_2), \tag{9.3}$$

where the last term of the Hamiltonian represents the Coulombian interaction between those two electrons.

The wave functions of this model are

$$|n\rangle = c_1|LL\rangle + c_2|LR\rangle + c_3|RL\rangle + c_4|RR\rangle. \tag{9.4}$$

This wave function has, for instance, a probability $|c_3|^2$ to find the electron number 1 close to the right-hand atom (R) and the electron number 2 close to the left-hand atom

Fundamentals of Magnetism. http://dx.doi.org/10.1016/B978-0-12-405545-2.00009-6

(L). It is not difficult to see that the Hamiltonian of this system is a 4×4 matrix and due to (i) those two atoms be equivalent and (ii) the Hamiltonian be symmetric in what concerns the exchange in position of the electrons, then several elements are equal; as, for instance:

$$U = \langle LL|V_c|LL\rangle = \langle RR|V_c|RR\rangle > 0 \tag{9.5}$$

and

$$J_D = \langle LR|V_c|RL\rangle = \langle RL|V_c|LR\rangle = \cdots > 0, \tag{9.6}$$

where U is the well-known Coulomb integral and describes the electronic repulsion between those two electrons in the same atom; while J_D is the well-known exchange integral and rules the overlap of the wave functions of the two electrons (and has no classical analogy). Note these two quantities are always positive. Thus, the Hamiltonian can be written as

$$\mathcal{H} = 2E_0 + \begin{pmatrix} U & t & t & J_D \\ t & 0 & J_D & t \\ t & J_D & 0 & t \\ J_D & t & t & U \end{pmatrix}, \tag{9.7}$$

where E_0 and t can be defined from the model of two atoms and one electron and are, respectively, the atomic energy (a constant for our purpose), and

$$t = \langle L|T|R\rangle \tag{9.8}$$

the hopping integral. This last rules the electron kinetics between two atoms ($T = -\hbar^2\nabla^2/2m$).

Now, we need to make the Hamiltonian diagonal. It is a simple task and after that we obtain as eigenvalues

$$E_1 = 2E_0 - J_D, \tag{9.9}$$
$$E_2 = 2E_0 + U - J_D, \tag{9.10}$$
$$E_3 = 2E_0 + \frac{U}{2} + J_D - \sqrt{4t^2 + \frac{U^2}{4}}, \tag{9.11}$$
$$E_4 = 2E_0 + \frac{U}{2} + J_D + \sqrt{4t^2 + \frac{U^2}{4}} \tag{9.12}$$

and wave functions

$$|1\rangle = \frac{1}{\sqrt{2}}(|LR\rangle - |RL\rangle), \tag{9.13}$$
$$|2\rangle = \frac{1}{\sqrt{2}}(|LL\rangle - |RR\rangle), \tag{9.14}$$
$$|3\rangle = \frac{\sin\vartheta}{\sqrt{2}}(|LL\rangle + |RR\rangle) + \frac{\cos\vartheta}{\sqrt{2}}(|LR\rangle + |RL\rangle), \tag{9.15}$$

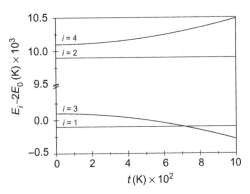

Figure 9.1 Energy spectra for the two atoms-two electrons model, as a function of the hopping t (expressed in Kelvin—K). $J_D = 10^2$ K and $U = 10^4$ K.

$$|4\rangle = \frac{\cos \vartheta}{\sqrt{2}}(|LL\rangle + |RR\rangle) - \frac{\sin \vartheta}{\sqrt{2}}(|LR\rangle + |RL\rangle), \tag{9.16}$$

where $\tan(2\vartheta) = -4t/U$.

From the spatial wave functions presented above, only $|1\rangle$ is antisymmetric; while the other three are symmetric.[1] However, the Pauli exclusion principle states that symmetric (antisymmetric) spatial wave functions lead to antiparallel (parallel) alignment of the spins. As an exercise, this fact is left to the reader to prove. We conclude thus that only the state $|1\rangle$ has a parallel alignment of the spins ("ferromagnetic"), and the remaining are antiparallel ("antiferromagnetic"). Note the ferro- and antiferromagnetic terms above are between quotes, because this simple model with two electrons does not provide a genuine ordering, i.e., there is no spontaneous ordering. Thus, the correct word is parallel and antiparallel, as used above.

In terms of order of magnitude, $U \gg J_D$ and then, as a function of hopping, the two levels of lower energy are E_1 (parallel alignment) and E_3 (antiparallel alignment). Figure 9.1 helps to visualize these energy spectra. We define then the gap between the parallel and antiparallel states as

$$J = (E_3 - E_1) = 2J_D + \frac{U}{2} - \sqrt{4t^2 + \frac{U^2}{4}} \tag{9.19}$$

known as exchange parameter.

---

[1] Remembering the reader, the antisymmetric wave function must change the sign when there is an exchange in position of the particles, i.e.,

$$\Psi(1, 2) = -\Psi(2, 1). \tag{9.17}$$

Analogously, the symmetric wave function must satisfy

$$\Psi(1, 2) = +\Psi(2, 1). \tag{9.18}$$

Thus, by construction, if $J > 0$ (and, consequently, $E_3 > E_1$), the ground state is $|1\rangle$, i.e., parallel spin alignment. Otherwise, if $J < 0$, E_3 is the lower energy and the ground state is $|3\rangle$, i.e., antiparallel spin alignment. We can also observe that the hopping integral t is responsible for the possible negative value of J, i.e., antiparallel alignment of the spins. Thus, the tendency to "ferromagnetism" decreases due to the increase of the hopping t.

9.1.2   Heisenberg model

This Hamiltonian considers localized wave functions, i.e., the hopping integral t is small. Thus, metals are not well described with this Hamiltonian; only *insulators*. For this case, the exchange parameter J can be rewritten as

$$J \simeq 2J_D - 4\frac{t^2}{U}. \tag{9.20}$$

This fact ratifies the points discussed above: J_D is always positive and rules the ferromagnetic behavior of the system; while the hopping integral t deviates the system to the antiferromagnetic arrangement (remember that $J > 0$ means parallel alignment of spins and $J < 0$ antiparallel).

Consider the interaction between two spins $\vec{S}_1$ and $\vec{S}_2$ as[2]

$$\mathcal{H}_{\mathrm{hei}} = -J\vec{S}_1 \cdot \vec{S}_2. \tag{9.21}$$

It is the famous Heisenberg Hamiltonian. We know that

$$(\vec{S}_1 + \vec{S}_2)^2 = \vec{S}^2 = \vec{S}_1^2 + \vec{S}_2^2 + 2\vec{S}_1 \cdot \vec{S}_2 \tag{9.22}$$

and then the eigenvalues of energy are

$$E_s = -\frac{J}{2}[s(s+1) - s_1(s_1+1) - s_2(s_2+1)]. \tag{9.23}$$

For two 1/2 spins there are only two possible values for s (it ranges from $|s_1 - s_2| \leq s \leq s_1 + s_2$): 0 (singlet—antiparallel alignment) and 1 (triplet—parallel alignment). Thus,

$$E_{s=0} = \frac{3}{4}J, \quad E_{s=1} = -\frac{1}{4}J. \tag{9.24}$$

The difference of energy between these two states is

$$E_{s=0} - E_{s=1} = J \tag{9.25}$$

in full accordance with Eq. (9.19). It justifies thus the way in which the Heisenberg Hamiltonian is written (see Eq. (9.21)).

---

[2] Note it is a spin Hamiltonian, but, of course, can be written for the total angular momentum $\vec{J}$.

9.1.3   Ising and XY models

The Heisenberg Hamiltonian above defined can be rewritten as below:

$$\mathcal{H}_{\text{hei}} = -J\vec{S}_1 \cdot \vec{S}_2 = -J(S_{1x}S_{2x} + S_{1y}S_{2y} + S_{1z}S_{2z}) \tag{9.26}$$

and then generalized as

$$\mathcal{H}_{\text{hei}} = -J[\alpha(S_{1x}S_{2x} + S_{1y}S_{2y}) + \beta S_{1z}S_{2z}]. \tag{9.27}$$

Note:

- $\alpha = \beta = 1$, recovers the isotropic Heisenberg term,
- $\alpha \neq \beta \neq 0$, anisotropic Heisenberg,
- $\alpha = 1$ and $\beta = 0$, XY Hamiltonian, and
- $\alpha = 0$ and $\beta = 1$, Ising Hamiltonian.

9.1.4   Bi-quadratic interaction

The isotropic interaction between spins can be extended up to higher-order terms of $\vec{S}_1 \cdot \vec{S}_2$. This is the aim of the bi-quadratic interaction

$$\mathcal{H}_{\text{biq}} = -j\left(\vec{S}_1 \cdot \vec{S}_2\right)^2. \tag{9.28}$$

9.1.5   Hubbard model

This model considers $J_D = 0$ and therefore the exchange parameter is

$$J = (E_3 - E_1) = \frac{U}{2} - \sqrt{4t^2 + \frac{U^2}{4}}. \tag{9.29}$$

Note, for the case $U \gg t$, this parameter assumes

$$J \simeq -4\frac{t^2}{U} \tag{9.30}$$

and, for $U \ll t$:

$$J \simeq \frac{U}{2} - 2t. \tag{9.31}$$

This model is used for both, metallic and insulating materials, depending on the value of U and t.

9.1.6   Antisymmetric: Dzialoshinsky–Moriya interaction

This kind of interaction arises for crystals of low symmetry, and it is not often to find. The spin Hamiltonian is given by

$$\mathcal{H}_{\text{anti}} = \vec{d} \cdot (\vec{S}_1 \times \vec{S}_2) \tag{9.32}$$

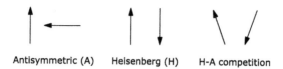

Figure 9.2 Left: Antisymmetric Dzialoshinsky–Moriya interaction. Center: Heisenberg isotropic interaction. Right: Competition of these two interactions that leads to a spin canted structure.

and was initially introduced by Dzialoshinsky and then detailed by Moriya [5,6].

This interaction tends to order the neighbor spins perpendicularly to each other; while the isotropic Heisenberg interaction tends to order those in either parallel or antiparallel fashion. Thus, this term is responsible for a spin canting structure. See Figure 9.2 for a better understanding.

9.1.7 Asymmetric: Dipolar Hamiltonian

Dipolar interaction arises due to the influence of the magnetic field created by one magnetic moment in other magnetic moment. Classically, the interaction energy is

$$E = \frac{1}{r^3}\left[\vec{\mu}_1 \cdot \vec{\mu}_2 - 3\frac{(\vec{\mu}_1 \cdot \vec{r})\,(\vec{\mu}_2 \cdot \vec{r})}{r^2}\right],\tag{9.33}$$

where $\vec{r}$ is the vector between those two magnetic moments, $\vec{\mu}_1$ and $\vec{\mu}_2$. From this energy, it is possible to write the Hamiltonian. Considering first

$$\vec{\mu}_S = -g_S \frac{\mu_B}{\hbar}\vec{S},\tag{9.34}$$

then

$$\mathcal{H}_d = \vec{S}_1 \cdot \overleftrightarrow{\mathcal{D}}_d \cdot \vec{S}_2 \,,\tag{9.35}$$

where $\overleftrightarrow{\mathcal{D}}$ is a tensor

$$\mathcal{D}_d^{(ab)} = (g\mu_B)^2 \left\langle \frac{r^2 \delta_{ab} - 3ab}{r^5}\right\rangle\tag{9.36}$$

and $\langle\cdots\rangle$ represents an integration with the spatial wave function. Note $g_S = g$. This Hamiltonian is named asymmetric and will be further discussed at the end of this chapter.

9.1.8 Mean field approach

Mean field theory assumes that the interaction of spins does not depend on the distance between those. It means that a certain spin interacts with its first neighbor equally as it interacts with another, even if placed in the opposite side of the crystal.

Well, one particle in a one-dimensional lattice has two first neighbors. Analogously, one particle in a two-dimensional square lattice has four first neighbors and in a three-dimensional cubic lattice, it has six first neighbors. Following this idea, since within the mean field theory one particle interacts with the same strength with all of the particles, this model is considered to describe a lattice with infinity dimension.

Thus, the magnetic field of induction $\vec{B}$ that one particle feels is

$$B = B_0 + \lambda M ,\tag{9.37}$$

where $\vec{B_0}$ is proportional to the applied magnetic field H_0 (remember, $B = \mu_0 H$), $\vec{M}$ is the total magnetization of the system, and λ is the mean field parameter that rules the strength of the interaction.

This interaction is the most important to the purpose of this book, since it will be used to describe the cooperative orderings: ferromagnetism, antiferromagnetism, and ferrimagnetism. Obviously, there are other ways to obtain a cooperative ordering, such as from the Heisenberg Hamiltonian; but these more elaborated models will not be described in this book.

9.2 Indirect exchange

In several materials, the interaction between two magnetic ions occurs directly, as described in the previous section. However, in some cases, the magnetic ions interact mediated by other ions or conducting electrons. These cases will be further discussed in this section.

9.2.1 Superexchange: oxides

Let us consider first two d orbitals linked by a p orbital. Note thus Figure 9.3-left: the orbitals d and p are disposed linearly ($180°$), maximizing the overlap of the wave functions d–p and p–d; favoring, therefore, the hopping. In accordance with the Pauli principle, there are some possible spin configurations, as presented in Figure 9.3-left-a–c. These cases are antiferromagnetic and ruled by the term

$$-4\frac{t^2}{U}\tag{9.38}$$

of Eq. (9.20).

For the other case (Figure 9.3-right), in which those d and p orbitals are orthogonal ($90°$), there is no overlap of the wave functions d–p and p–d and, due to the Pauli principle, there is only one possible configuration; as shown in Figure 9.3-right-d. Note, due to the low hopping, this configuration must be ferromagnetic.

These ideas can be summarized by the Goodenough–Kanamori rules, in which it states that the arrangement between magnetic d ion bridged by p ions depends on the angle of the d–p–d path. Summarizing:

$$180° \Rightarrow \text{maximum overlap} \Rightarrow \text{hopping} \Rightarrow \text{antiferromagnetism},\tag{9.39}$$

$$90° \Rightarrow \text{minimum overlap} \Rightarrow \text{zero hopping} \Rightarrow \text{ferromagnetism}.\tag{9.40}$$

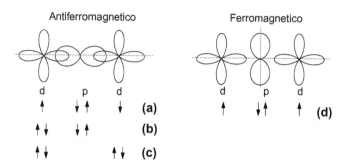

Figure 9.3 Schematic representation of the Goodenough–Kanamori rules. Left: Possible anti-ferromagnetic arrangement, in accordance with the Pauli principle and electron mobility due to the overlap of the wave functions. Right: Ferromagnetic configuration, in accordance with the Pauli principle. Note this last case has a lower mobility, since there is no overlap of the wave functions.

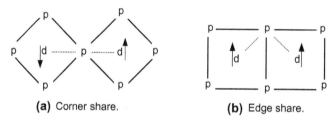

Figure 9.4 Top view of p octahedral environment with a central d ion. Note that a corner share leads to antiparallel alignment of the d ions; while an edge share leads to a parallel alignment. See text and Figure 9.3 for a better understanding.

This rule can be seen in the other way. Suppose a d ion inside an octahedral p environment. If two neighboring octahedra edge share an equatorial plane, it corresponds to the case of Figure 9.3-right (i.e., a ferromagnetic case). On the contrary, if those two neighboring octahedra corner share its equatorial plan, this situation is given by Figure 9.3-left (i.e., an antiferromagnetic case). Figure 9.4 clarifies these arguments.

9.2.2 RKKY model: metals

This interesting model was developed by Ruderman–Kittel–Kasuya–Yosida and considers that localized magnetic moments interact mediated by a conducting electron. A complete description of this model is not trivial and then is out of the scope of this book. However, we can present a simple description; the roots to guide the ideas. In few words, we will obtain an oscillatory function that rules the interaction between these two localized spins (mediated by an itinerant electron). However, to obtain it, we need first to write the probability to find an electron with spin either up or down. Meanwhile, to obtain this information, we need to write the wave function of these electrons via perturbation theory, that, on its turn, depends on the Hamiltonian.

Let us consider a Heisenberg-like interaction between a localized spin $\vec{S}$ and an itinerant spin $\vec{\mathcal{S}}$

$$\mathcal{H} = -J(\vec{r})\vec{S} \cdot \vec{\mathcal{S}}, \tag{9.41}$$

where $J(\vec{r})$ means that the strength of this interaction depends on the distance of those two spins. From Eq. (3.12), we can rewrite the above Hamiltonian in terms of ladder operators

$$\mathcal{H} = -J(\vec{r})\left[\frac{1}{2}\left(S_+\mathcal{S}_- + S_-\mathcal{S}_+\right) + S_z\mathcal{S}_z\right]. \tag{9.42}$$

This system has, at least, two important parameters: one related to the z projection of the localized spin (say, m_s), and the other associated to the kinetics of the itinerant electron (say, k). Thus, we can define an operator

$$\Psi(\vec{r}) = \sum_{k,m_s} e^{i\vec{k}\cdot\vec{r}} |m_s\rangle c_{k,m_s} \tag{9.43}$$

and its adjoint

$$\Psi^\dagger(\vec{r}) = \sum_{k,m_s} c^\dagger_{k,m_s} \langle m_s| e^{-i\vec{k}\cdot\vec{r}}. \tag{9.44}$$

Note we are considering a plane wave like for the itinerant electron. In addition, $c^\dagger_{k,m_s}$ and c_{k,m_s} mean creation and annihilation operators, respectively.

Consider now the transformation below

$$H = \int \Psi^\dagger(\vec{r})\mathcal{H}\Psi(\vec{r})\mathrm{d}\vec{r}$$

$$= \sum_{k,m_s}\sum_{k',m'_s} c^\dagger_{k',m'_s} c_{k,m_s} J(\vec{k},\vec{k}')\langle m'_s|\left[\frac{1}{2}\left(S_+\mathcal{S}_- + S_-\mathcal{S}_+\right) + S_z\mathcal{S}_z\right]|m_s\rangle, \tag{9.45}$$

where

$$J(\vec{k},\vec{k}') = \int J(\vec{r})e^{i(\vec{k}-\vec{k}')\cdot\vec{r}}\mathrm{d}\vec{r}. \tag{9.46}$$

In spite of it seeming to be a mean number, H is still an operator, however, it depends on the creation and annihilation operators, instead of those initial spin operators. We know that $|m_s\rangle$ (of $s = 1/2$) can assume two values: $|\uparrow\rangle$ and $|\downarrow\rangle$; and the eigenvalues of above are:

$$\mathcal{S}_+|\uparrow\rangle = 0 \quad \mathcal{S}_-|\uparrow\rangle = |\downarrow\rangle \quad \mathcal{S}_z|\uparrow\rangle = \frac{1}{2}|\uparrow\rangle,$$

$$\mathcal{S}_+|\downarrow\rangle = |\uparrow\rangle \quad \mathcal{S}_-|\downarrow\rangle = 0 \quad \mathcal{S}_z|\downarrow\rangle = -\frac{1}{2}|\downarrow\rangle. \tag{9.47}$$

Thus, the second quantization Hamiltonian is

$$H = -\frac{J_0}{2} \sum_{k,k'} \left[S_z \left(c^\dagger_{k',\uparrow} c_{k,\uparrow} - c^\dagger_{k',\downarrow} c_{k,\downarrow} \right) + S_+ c^\dagger_{k',\downarrow} c_{k,\uparrow} + S_- c^\dagger_{k',\uparrow} c_{k,\downarrow} \right],$$

(9.48)

where we considered $J(\vec{r}) = J_0 \delta(\vec{r})$ and therefore $J(\vec{k}, \vec{k}') = J_0$, i.e., a constant.

From the perturbation theory (see Complement 8.4), we know that the wave function (up to the first-order correction) is

$$|k, \uparrow\rangle = |k, \uparrow^{(0)}\rangle + \sum_{k \neq k', m_s} \frac{\langle^{(0)} k', m_s | H | k, \uparrow^{(0)}\rangle}{\frac{\hbar^2}{2m} (k^2 - k'^2)} |k', m_s^{(0)}\rangle.$$

(9.49)

Remember, c_{k,m_s} annihilates an electron with $|k, m_s\rangle$ state; while $c^\dagger_{k,m_s}$ creates an electron at that state. These operators must follow the rules below (easy to see):

$$c^\dagger_{k',\uparrow} c_{k,\uparrow} |\uparrow\rangle = |\uparrow\rangle, \quad c^\dagger_{k',\uparrow} c_{k,\uparrow} |\downarrow\rangle = 0,$$

$$c^\dagger_{k',\downarrow} c_{k,\downarrow} |\uparrow\rangle = 0, \quad c^\dagger_{k',\downarrow} c_{k,\downarrow} |\downarrow\rangle = |\downarrow\rangle,$$

$$c^\dagger_{k',\downarrow} c_{k,\uparrow} |\uparrow\rangle = |\downarrow\rangle, \quad c^\dagger_{k',\downarrow} c_{k,\uparrow} |\downarrow\rangle = 0,$$

$$c^\dagger_{k',\uparrow} c_{k,\downarrow} |\uparrow\rangle = 0, \quad c^\dagger_{k',\uparrow} c_{k,\downarrow} |\downarrow\rangle = |\uparrow\rangle,$$

(9.50)

and therefore, after evaluation of the $\langle^{(0)} k', m_s | H | k, \uparrow^{(0)}\rangle$ sandwich, we get

$$|k, \uparrow\rangle = |k, \uparrow^{(0)}\rangle + \frac{m J_0}{\hbar^2} \sum_{k \neq k'} \frac{S_+ |k', \downarrow^{(0)}\rangle + S_z |k', \uparrow^{(0)}\rangle}{k'^2 - k^2}.$$

(9.51)

Considering huge reciprocal spaces (as done before, Chapter 6), we can change the above sum to an integral, following

$$\sum_{k'} \rightarrow \frac{V}{(2\pi)^3} \int d\vec{k}'$$

(9.52)

and then, the second term of Eq. (9.51) reads as

$$\frac{m J_0}{\hbar^2} \frac{V}{(2\pi)^3} \left[S_+ |\downarrow^{(0)}\rangle + S_z |\uparrow^{(0)}\rangle \right] \int \frac{e^{i\vec{k}'\cdot\vec{r}}}{k'^2 - k^2} d\vec{k}'.$$

(9.53)

The above integral is well known[3] and therefore the wave function can be rewritten as

$$|k, \uparrow\rangle = |k, \uparrow^{(0)}\rangle + \frac{m J_0}{\hbar^2} \frac{V}{8\pi^2} \frac{\cos(kr)}{r} \left[S_+ |\downarrow^{(0)}\rangle + S_z |\uparrow^{(0)}\rangle \right].$$

(9.55)

[3]
$$\int \frac{e^{i\vec{k}'\cdot\vec{r}}}{k'^2 - k^2} d\vec{k}' = \pi \frac{\cos(kr)}{r}.$$

(9.54)

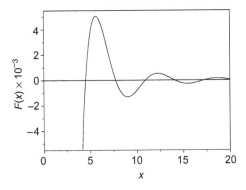

Figure 9.5 Behavior of $F(x)$ function from the RKKY model (see Eq. (9.59)).

Now we are able to write the probability to find an electron with up projection (considering $|k\rangle = e^{i\vec{k}\cdot\vec{r}}$)

$$P(k,\uparrow) = \langle k,\uparrow|k,\uparrow\rangle = 1 + \frac{m J_0 V}{h^2} \frac{\cos(kr)\cos(\vec{k}\cdot\vec{r})}{r} S_z \qquad (9.56)$$

and then the probability to find an electron with up projection and any k state is

$$P(\uparrow) = \frac{V}{(2\pi)^3} \int P(k,\uparrow)d\vec{k} \qquad (9.57)$$

$$= \frac{V}{8\pi^3} 2\pi \int_0^{k_F} \int_0^{\pi} \left(1 + \frac{m J_0 V}{h^2} \frac{\cos(kr)\cos(kr\cos\theta)}{r} S_z\right) k^2 \sin\theta d\theta\, dk.$$

After a few more simple steps we obtain

$$P(\uparrow) = \frac{V}{6\pi^2} k_F^3 - \frac{m J_0 V^2 k_F^4}{\pi^2 h^2} S_z F(2k_F r), \qquad (9.58)$$

where

$$F(x) = \frac{x \cos x - \sin x}{x^4}. \qquad (9.59)$$

The above function is our objective (see Figure 9.5). Note therefore that the probability to find an electron with up projection depends on the distance between the itinerant electron and the localized one, following an oscillatory fashion, where the period depends on the Fermi energy of the system.

The results above were obtained for one itinerant electron and one localized spin. To go further into the model, we need to consider two localized spins and one itinerant electron (that mediates the interaction). Thus, the new Hamiltonian reads as

$$\mathcal{H} = -\left[J_1(\vec{r})\vec{S}_1 + J_2(\vec{r} - \vec{R})\vec{S}_2\right] \cdot \vec{\mathcal{S}}. \qquad (9.60)$$

However, the development of this new model is quite huge to be described here; but the steps are similar to those above presented. The most important result is below. From perturbation theory (see Complement 8.A), the second-order correction to the above Hamiltonian is

$$\mathcal{H}_{RKKY} = -J(r)\vec{S}_1 \cdot \vec{S}_2, \qquad (9.61)$$

where

$$J(r) = \mathcal{J}F(2k_Fr) \qquad (9.62)$$

and $\mathcal{J}$ is a positive constant. It means that one localized electron interacts with other mediated by an itinerant electron; and this interaction is Heisenberg-like. However, the strength $\mathcal{J}F(2k_Fr)$ of this interaction depends on the distance between these two spins and oscillates between ferromagnetic (positive strength) and antiferromagnetic (negative strength).

9.3 Asymmetric: Spin–orbit coupling

The spin–orbit coupling arises from the interaction of the spin with the magnetic field from its own orbit. Classically, this energy is

$$E = -\vec{\mu}_s \cdot \vec{H}_{\text{orb}}. \qquad (9.63)$$

Well, the magnetic field at the center of a ring of radius a is (see Section 1.1)

$$\vec{H}_{\text{orb}} = \frac{e}{4\pi ma^3}\vec{L}, \qquad (9.64)$$

where $\vec{L}$ is the angular momentum due to the electron precession. We also know that the magnetic moment $\vec{\mu}_s$ due to a spin $\vec{S}$ is (Eq. (3.37))

$$\vec{\mu}_s = -g_S\frac{\mu_B}{\hbar}\vec{S}. \qquad (9.65)$$

Thus, the classical spin–orbit coupling energy can be written as

$$E = \left(g_S\frac{\mu_B}{\hbar}\right)\left(\frac{e}{4\pi ma^3}\right)\vec{S} \cdot \vec{L}. \qquad (9.66)$$

This energy represents the cost to remove the spin from a direction parallel to the angular momentum, as represented in Figure 3.1. If we consider now that this system belongs to a crystal lattice and the angular momentum is coupled to this lattice, we have thus an anisotropic system. In other words, if we remove the spin from its direction of low energy (parallel to the orbital momentum and then, also coupled to the lattice), and lead it to another direction, this procedure consumes energy and this energy is known as local magnetocrystalline anisotropy. It is important to mention that this energy is of the order of 10^1 K for 3d ions.

Let us now go further discussing the spin–orbit coupling, recovering the quantum-relativistic result (from the Dirac equation—see Chapter 2)

$$\mathcal{H}_{SL} = \zeta \vec{S} \cdot \vec{L},$$ (9.67)

where ζ is the spin–orbit parameter. Remember, this parameter is positive when the shell is less than half-filled; and negative when the shell is more than half-filled. Note the above equation is similar to the classical one (Eq. (9.66)), however, the pre-factor is different (compare Eqs.(2.69) and (9.66)).

Consider this Hamiltonian as a perturbation (see Complement 8.A); thus, the second-order correction is given by

$$\mathcal{H}_{\text{ani}} = \zeta^2 \sum_{k \neq n} \frac{\langle k | \vec{L} \cdot \vec{S} | n \rangle \langle n | \vec{L} \cdot \vec{S} | k \rangle}{\varepsilon_n - \varepsilon_k},$$ (9.68)

that can be rewritten as

$$\mathcal{H}_{\text{ani}} = \vec{S} \cdot \overset{\leftrightarrow}{\mathcal{D}}_{\text{ani}} \cdot \vec{S},$$ (9.69)

where

$$\mathcal{D}_{\text{ani}}^{(ab)} = \zeta^2 \sum_{k \neq n} \frac{\langle k | L_a | n \rangle \langle n | L_B | k \rangle}{\varepsilon_n - \varepsilon_k}.$$ (9.70)

Note the above Hamiltonian describes the second-order correction to the energy, considering the spin–orbit coupling as a perturbation. This correction $\mathcal{H}_{\text{ani}}$ represents the local magnetocrystalline anisotropy and is also known as asymmetric, due to the asymmetry of the $\overset{\leftrightarrow}{\mathcal{D}}_{\text{ani}}$ tensor.

9.4 Zeeman interaction

Analogously to before, the Zeeman interaction has its complete form as described on the Dirac equation (see Chapter 2)

$$\mathcal{H}_{\text{zee}} = \frac{\mu_B}{\hbar} \left(\vec{L} + 2\vec{S} \right) \cdot \vec{B}$$ (9.71)

and, from Chapter 3, it can be rewritten as

$$\mathcal{H}_{\text{zee}} = g \frac{\mu_B}{\hbar} \vec{J} \cdot \vec{B}$$ (9.72)

This Hamiltonian refers thus to the interaction of the magnetic moments with the external magnetic field of induction $\vec{B}$.

9.4.1   The g tensor

The Landé factor is not a simple number and this section discusses this point. Let us start considering the Zeeman interaction

$$\mathcal{H}_{zee} = \mu_B \vec{B} \cdot \left(\vec{L} + 2\vec{S} \right) \tag{9.73}$$

and the spin–orbit coupling

$$\mathcal{H}_{SL} = \zeta \vec{S} \cdot \vec{L}. \tag{9.74}$$

Now, we need to change $\mathcal{H}_{zee} + \mathcal{H}_{SL}$ to a spin Hamiltonian form, i.e., rewrite the Hamiltonian in such way as to collect in a unique variable all of orbital dependence, remaining therefore only the spin operators. Thus

$$\mathcal{H} = \mathcal{H}_{zee} + \mathcal{H}_{SL} = 2\mu_B \vec{S} \cdot \vec{B} + \mu_B \vec{L} \cdot \vec{B} + \zeta \vec{S} \cdot \vec{L}. \tag{9.75}$$

The first term does not depend on the orbital moment and then it is already a spin Hamiltonian. Let us work therefore on the basis in which this first term is diagonal and consider the following two terms as a perturbation. Thus, considering the second-order perturbation theory (see Complement 8.A), we have

$$\mathcal{H} = 2\mu_B \vec{S} \cdot \vec{B} + \sum_{k \neq n} \frac{|\langle k | \mu_B \vec{L} \cdot \vec{B} + \zeta \vec{S} \cdot \vec{L} | n \rangle|^2}{\varepsilon_n - \varepsilon_k}, \tag{9.76}$$

where the $|i\rangle$ states above used are those that diagonalize the first term of the Hamiltonian. Opening these terms we get

$$\mathcal{H} = +\mu_B^2 \vec{B} \cdot \overleftrightarrow{\mathcal{D}}_L \cdot \vec{B} + \zeta^2 \vec{S} \cdot \overleftrightarrow{\mathcal{D}}_L \cdot \vec{S} + \mu_B \vec{B} \cdot \overleftrightarrow{g} \cdot \vec{S}, \tag{9.77}$$

where

$$\mathcal{D}_L^{(ab)} = \sum_{k \neq n} \frac{\langle k | L_a | n \rangle \langle n | L_B | k \rangle}{\varepsilon_n - \varepsilon_k} \tag{9.78}$$

and

$$g^{(ab)} = 2 \left(\mathbb{I} + \zeta \mathcal{D}_L^{(ab)} \right). \tag{9.79}$$

Note $\overleftrightarrow{\mathcal{A}}$ and $\mathcal{A}^{(ab)}$ are different ways to write a rank-2 tensor.

The second term of the above Hamiltonian is that obtained discussing the spin–orbit coupling, i.e., the local magnetocrystalline anisotropy; and, comparing with Eq. (9.69), we obtain

$$\overleftrightarrow{\mathcal{D}}_L = \frac{1}{\zeta^2} \overleftrightarrow{\mathcal{D}}_{ani}. \tag{9.80}$$

The third term represents the Zeeman interaction

$$\mathcal{H}_{zee} = \mu_B \vec{B} \cdot \overleftrightarrow{g} \cdot \vec{S}. \tag{9.81}$$

Note for this spin Hamiltonian $g = 2$ and any deviation from this value is due to an orbital contribution.

9.5   The $\mathcal{D}$ tensor

Let us now describe in more detail the $\mathcal{D}$ tensor, starting from the asymmetric Hamiltonian

$$\mathcal{H} = \vec{S}_i \cdot \overleftrightarrow{\mathcal{D}} \cdot \vec{S}_j = \sum_u \sum_v \mathcal{D}_{uv} S_{iu} S_{jv}, \tag{9.82}$$

where $(u, v) = x, y, z$. Note that, if $i = j$, this asymmetric case represents the local magnetocrystalline anisotropy; while, if $i \neq j$, it is the dipolar case. Considering that the axis of the system is the same as the $\overleftrightarrow{\mathcal{D}}$ tensor, only the diagonal elements are non zero. Thus

$$\mathcal{H} = \mathcal{D}_{xx} S_{ix} S_{jx} + \mathcal{D}_{yy} S_{iy} S_{jy} + \mathcal{D}_{zz} S_{iz} S_{jz}. \tag{9.83}$$

For convenience, let us define three parameters associated to D_{uv}

$$D = \frac{1}{2}(-\mathcal{D}_{xx} - \mathcal{D}_{yy} + 2\mathcal{D}_{zz}), \tag{9.84}$$

$$E = \frac{1}{2}(\mathcal{D}_{xx} - \mathcal{D}_{yy}), \tag{9.85}$$

$$K = \frac{1}{3}(\mathcal{D}_{xx} + \mathcal{D}_{yy} + \mathcal{D}_{zz}), \tag{9.86}$$

where we can write this linear system of equation in the matrix form

$$\begin{pmatrix} D \\ E \\ K \end{pmatrix} = \mathbb{T} \begin{pmatrix} \mathcal{D}_{xx} \\ \mathcal{D}_{yy} \\ \mathcal{D}_{zz} \end{pmatrix}, \quad \text{where } \mathbb{T} = \begin{pmatrix} -1/2 & -1/2 & 1 \\ 1/2 & -1/2 & 0 \\ 1/3 & 1/3 & 1/3 \end{pmatrix} \tag{9.87}$$

and, consequently

$$\begin{pmatrix} \mathcal{D}_{xx} \\ \mathcal{D}_{yy} \\ \mathcal{D}_{zz} \end{pmatrix} = \mathbb{T}^{-1} \begin{pmatrix} D \\ E \\ K \end{pmatrix}, \quad \text{where } \mathbb{T}^{-1} = \begin{pmatrix} -1/3 & 1 & 1 \\ -1/3 & -1 & 1 \\ 2/3 & 0 & 1 \end{pmatrix}. \tag{9.88}$$

Now, we have the elements of the diagonal of the tensor $\overleftrightarrow{\mathcal{D}}$ as a function of the parameters above defined

$$\mathcal{D}_{xx} = -\frac{1}{3}D + E + K, \tag{9.89}$$

$$\mathcal{D}_{yy} = -\frac{1}{3}D - E + K, \tag{9.90}$$

$$\mathcal{D}_{zz} = \frac{2}{3}D + K. \tag{9.91}$$

It is now possible to rewrite the asymmetric Hamiltonian, however, as a function of the new parameters

$$\mathcal{H} = D\left(S_{iz}S_{jz} - \frac{1}{3}\vec{S}_i \cdot \vec{S}_j\right) + E\left(S_{ix}S_{jx} - S_{iy}S_{jy}\right) + K(\vec{S}_i \cdot \vec{S}_j). \tag{9.92}$$

To ensure that the trace of the tensor $\overset{\leftrightarrow}{\mathcal{D}}$ is null, we can assume $K = 0$. Then

$$D = \frac{3}{2}\mathcal{D}_{zz} = -\frac{3}{2}(\mathcal{D}_{xx} + \mathcal{D}_{yy}), \tag{9.93}$$

$$E = \frac{1}{2}(\mathcal{D}_{xx} - \mathcal{D}_{yy}) \tag{9.94}$$

and the Hamiltonian can be rewritten as

$$\mathcal{H} = D\left(S_{iz}S_{jz} - \frac{1}{3}\vec{S}_i \cdot \vec{S}_j\right) + E\left(S_{ix}S_{jx} - S_{iy}S_{jy}\right). \tag{9.95}$$

It is easy to note that the parameter d will always be on the principal diagonal of the Hamiltonian, since both $S_{iz}S_{jz}$ and $\vec{S}_i \cdot \vec{S}_j$ are diagonals on the basis $|s_i, s_j, s, m_s\rangle$; while the parameter E will always be on the off-diagonal, since S_x and S_y are not diagonal in this basis. Due to this fact, d is named as axial parameter and E as rhombic parameter.

The above results are applicable directly to the dipolar Hamiltonian, considering $i \neq j$. In what concerns the local magnetocrystalline anisotropy, when $i = j$, it is easy to see that

$$\mathcal{H} = D\left(S_z^2 - \frac{1}{3}S^2\right) + E\left(S_x^2 - S_y^2\right). \tag{9.96}$$

9.5.1   Coupling of $\mathcal{D}$ tensors: dipolar and local

Consider a dimer (i.e., only two coupled spins), with dipolar interaction and local magnetocrystalline anisotrpoy (s_1 and s_2 must be bigger than 1/2, otherwise the local magnetocrystalline anisotropy is zero—see exercise of this chapter). The Hamiltonian can be written as

$$\mathcal{H} = \vec{S}_1 \cdot \overset{\leftrightarrow}{\mathcal{D}}_1 \cdot \vec{S}_1 + \vec{S}_2 \cdot \overset{\leftrightarrow}{\mathcal{D}}_2 \cdot \vec{S}_2 + \vec{S}_1 \cdot \overset{\leftrightarrow}{\mathcal{D}}_{12} \cdot \vec{S}_2, \tag{9.97}$$

where the two first terms refer to the local anisotropy and the last to the dipolar interaction. This Hamiltonian can be rewritten as

$$\mathcal{H} = \vec{S} \cdot \overset{\leftrightarrow}{\mathcal{D}}_s \cdot \vec{S}, \tag{9.98}$$

where

$$\vec{S} = \vec{S}_1 + \vec{S}_2 \tag{9.99}$$

and

$$\overset{\leftrightarrow}{\mathcal{D}_s} = C_1\overset{\leftrightarrow}{\mathcal{D}_1} + C_2\overset{\leftrightarrow}{\mathcal{D}_2} + C_{12}\overset{\leftrightarrow}{\mathcal{D}_{12}}. \tag{9.100}$$

After some calculation (out of the scope of this book)

$$C_1 = \frac{1}{4}(1 + c_+) + \frac{1}{2}c_-, \tag{9.101}$$

$$C_2 = \frac{1}{4}(1 + c_+) - \frac{1}{2}c_-, \tag{9.102}$$

$$C_{12} = \frac{1}{4}(1 - c_+), \tag{9.103}$$

where

$$c_+ = 1 + \frac{6[s_1(s_1+1) - s_2(s_2+1)]^2 - 2s(s+1)[s(s+1) + 2s_1(s_1+1) + 2s_2(s_2+1)]}{(2s+3)(2s-1)s(s+1)} \tag{9.104}$$

and

$$c_- = \frac{[4s(s+1) - 3][s_1(s_1+1) - s_2(s_2+1)]}{(2s+3)(2s-1)s(s+1)}. \tag{9.105}$$

Note that s can assume values from $|s_1 - s_2|$ up to $|s_1 + s_2|$. For instance, for a $s = s_2 = 1$ dimer, the values of s are 2, 1, and 0.

10 Long-Range Ordering

The previous chapters present exhaustively the magnetic properties of localized and itinerant electrons with both, diamagnetic and paramagnetic character, i.e., noncooperative phenomena. To obtain the magnetic properties of cooperative systems, the magnetic interactions must be taken into account; for instance, considering the mean field approach. Thus, the results and discussions of the present chapter will be based on the mean field approach applied to the previous results of noncooperative systems.

Generally speaking, ferromagnetic arrangement has only one magnetic lattice and all of the magnetic moments (even if there is more than one type) are spontaneously aligned along a particular direction. On the other hand, antiferromagnetic arrangement needs two sublattices to be understood; one sublattice points in opposition to the other, and both have the same values of Lande factor, total angular momentum, and intra-sublattice magnetic interaction. Finally, the ferrimagnetic ordering also needs two sublattices, as the antiferromagnetic case, however, each sublattice has its own Lande factor, total angular momentum, and intra-sublattice interaction. Figure 10.1 is useful to see these sublattice contributions for each arrangement.

10.1 Ferromagnetism

10.1.1 Localized ferromagnetism

First, let us recover the localized paramagnetic magnetization, given by Eq. (8.51)

$$M(T, B, N) = Ngj\mu_B B_j(x),$$
(10.1)

where

$$B_j(x) = a \coth(ax) - b \coth(bx)$$
(10.2)

is the Brillouin function,

$$a = 1 + b \quad \text{and} \quad b = \frac{1}{2j}.$$
(10.3)

In addition,

$$x = \frac{gj\mu_B B}{k_B T}$$
(10.4)

Fundamentals of Magnetism. http://dx.doi.org/10.1016/B978-0-12-405545-2.00010-2

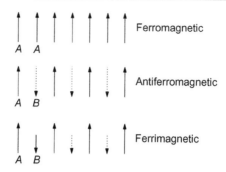

Figure 10.1 Ferro-, antiferro-, and ferrimagnetic (sub) lattice arrangements. A and B label each sublattice. Note the ferromagnetic case has only one lattice.

and, considering the mean field approximation, given by Eq. (9.37)

$$B = B_0 + \lambda M. \tag{10.5}$$

10.1.1.1   Magnetization

From the above, the ferromagnetic magnetization can be written as

$$M = N g j \mu_B B_j \left(\frac{g j \mu_B}{k_B T} [B_0 + \lambda M] \right). \tag{10.6}$$

Note the magnetization M depends on the magnetization that is into the argument of the Brillouin function and therefore there is no analytical solution for this quantity. Thus, to understand the behavior of this ordered case, a numerical evaluation is needed.

Magnetization as a function of temperature has a critical point, named Curie temperature T_c, below which there is a spontaneous magnetization M_{sp}, i.e., nonzero magnetization for zero applied magnetic field: $M_{sp} = M(T < T_c, B_0 = 0) \neq 0$. Above T_c, for zero external magnetic field, the magnetization is zero. Note T_c is only defined for $B_0 = 0$, otherwise the magnetization never goes to zero and this critical temperature cannot thus be defined. Figure 10.2a depicts these facts. Note for $b_0 = \mu_B B_0 / k_B T_c = 0$ the magnetization goes to zero at $T = T_c$, while for $b_0 \neq 0$ there is a tail, analogous to the paramagnetic case. From the experimental point of view, since it is not possible to measure magnetization without external magnetic field, the Curie temperature is obtained as the extreme point of the $\partial M / \partial T$ curve.

The magnetization as a function of the external applied magnetic field b_0 has a similar behavior of the paramagnetic case. However, for temperatures below T_c the magnetization is not zero for zero magnetic field. In addition, the external magnetic field needed to fully align (saturate) the magnetic moments of the specimen is much smaller than the paramagnetic case, since for the ferromagnetic case, in addition to the external magnetic field, there is a mean field λM under the magnetic moments. See Figure 10.2b for a better understanding.

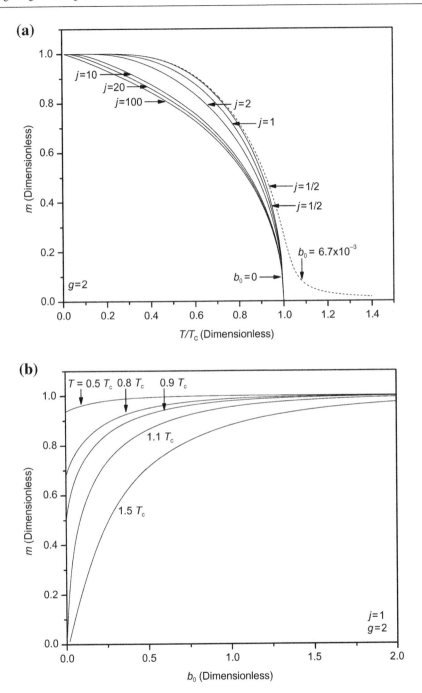

Figure 10.2 Ferromagnetic reduced magnetization $m = M/N\mu_B$ as a function of (a) reduced temperature T/T_c and (b) reduced external magnetic field $b_0 = \mu_B B_0/k_B T_c$. Results obtained numerically.

10.1.1.2   Curie temperature

To determine the Curie temperature T_c, the external applied magnetic field B_0 must be zero and, in addition, the magnetization must be small. Thus, we need to recover other result from paramagnetism: the limit of the Brillouin magnetization (given by Eq. (8.51)), however, for small values of B and, consequently, small values of magnetization. This result is given by Eq. (8.60) and is reproduced below

$$M = \frac{Ng^2\mu_B^2 j(j+1)}{3k_B T} B.$$
(10.7)

Considering the mean field approximation, $B_0 = 0$ and $T = T_c$ into the above equation we obtain

$$T_c = \frac{Ng^2\mu_B^2 j(j+1)\lambda}{3k_B}.$$
(10.8)

Note T_c is proportional to the mean field parameter λ, that rules the strength of the mean field interaction. It is important to stress that two parameters characterize a ferromagnetic material: T_c and the saturation value of the magnetization gj. In addition, for temperatures above T_c, the ferromagnetic system does not transit to a paramagnetic material, but only behaves like one. In other words, a simple ferromagnetic material is always a simple ferromagnetic material; for temperatures above T_c the thermal energy is bigger than the magnetic interactions (ruled by λ) and the system behaves like a paramagnetic material.

10.1.1.3   Magnetic susceptibility for $T > T_c$

The magnetic susceptibility above T_c is also an important quantity, since it gives information of "free" magnetic moments of the material. Note "free" means the magnetic interactions are smaller than the thermal energy of the temperature scale under consideration. To obtain the magnetic susceptibility, we need to go back to Eq. (10.7), however, considering $B_0 \neq 0$. Thus,

$$M = \frac{Ng^2\mu_B^2 j(j+1)}{3k_B T}(\mu_0 H_0 + \lambda M)$$
(10.9)

and then the famous Curie–Weiss susceptibility arises[1]

$$\chi = \frac{C}{T - \theta_p},$$
(10.11)

[1] Remember

$$\chi = \lim_{H_0 \to 0} \frac{\partial M}{\partial H_0}.$$
(10.10)

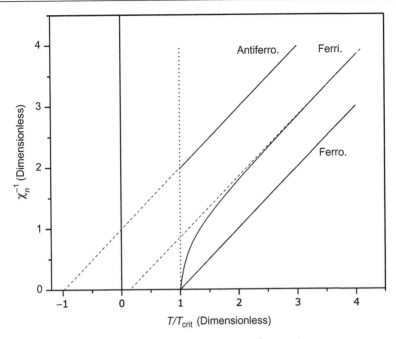

Figure 10.3 Inverse normalized magnetic susceptibility $\chi_n^{-1} = \chi^{-1}C/T_{\mathrm{crit}}$, where T_{crit} is the critical temperature of the system. Quantities obtained numerically. For the antiferromagnetic case: $\alpha = 0.01$. For the ferrimagnetic case: $\alpha = 0.01$ and $\beta = 10$.

where C is the Curie constant, given by Eq. (8.62) and reproduced below

$$C = \frac{N\mu_0 p_{\mathrm{eff}}^2}{3k_{\mathrm{B}}}. \tag{10.12}$$

Above

$$p_{\mathrm{eff}}^2 = g^2 j(j+1)\mu_{\mathrm{B}}^2 \tag{10.13}$$

is the paramagnetic effective moment, defined in Eq. (8.63). In addition

$$\theta_{\mathrm{p}} = \frac{C\lambda}{\mu_0} > 0 \tag{10.14}$$

is the paramagnetic Curie temperature. Note for this mean field model, $T_{\mathrm{c}} = \theta_{\mathrm{p}}$. Figure 10.3 presents the inverse magnetic susceptibility. The angular coefficient of this straight line is ruled by the effective paramagnetic moment p_{eff}, that measures the total angular momentum j and the Lande factor g.

10.1.2 Itinerant ferromagnetism

The paramagnetic behavior of itinerant electrons was described in Section 8.2 and, for the low temperature regime ($\epsilon_{\mathrm{F}} \gg k_{\mathrm{B}}T$), the reduced magnetization $m = M/N\mu_{\mathrm{B}}$

was given by Eq. (8.122), reproduced below

$$\mathcal{P} - \frac{\pi^2}{12} t^2 \mathcal{L} = 2b, \tag{10.15}$$

where $b = \mu_B B / \epsilon_{F0}$ and $t = k_B T / \epsilon_{F0}$ are, respectively, the reduced and dimensionless temperature and magnetic field. In addition,

$$\mathcal{P} = (1+m)^{2/3} - (1-m)^{2/3} \tag{10.16}$$

and

$$\mathcal{L} = (1+m)^{-2/3} - (1-m)^{-2/3}. \tag{10.17}$$

For small values of m and t, the above equation reads as (Eq. (8.124))

$$m = \frac{3}{2} b \left[1 - \frac{\pi^2}{12} t^2 \right]. \tag{10.18}$$

To describe the ferromagnetic behavior of these itinerant electrons, let us consider the mean field approach. Thus, the reduced magnetic field reads as

$$b = \frac{\mu_B B}{\epsilon_{F0}} = b_0 + \frac{2}{3} g(\epsilon_{F0}) U m, \tag{10.19}$$

where $b_0 = \mu_B B_0 / \epsilon_{F0}$ is the reduced applied magnetic field, $U = \mu_B^2 \lambda$ is known as *Stoner parameter* (better discussed below), and $g(\epsilon_{F0})$ is the density of states at the Fermi level, related to the Fermi level through Eq. (6.32), reproduced below

$$g(\epsilon_{F0}) = \frac{3}{2} \frac{N}{\epsilon_{F0}}. \tag{10.20}$$

10.1.2.1 Magnetization

Inserting Equation (10.19) into Eq. (10.15) leads to

$$b_0(t, m) = \frac{1}{2} \left(\mathcal{P} - \frac{\pi^2 t^2}{12} \mathcal{L} - \frac{4}{3} g(\epsilon_{F0}) U m \right) \tag{10.21}$$

or

$$t(b_0, m) = \left\{ \frac{12}{\pi^2 \mathcal{L}} \left[\mathcal{P} - 2b_0 - \frac{4}{3} g(\epsilon_{F0}) U m \right] \right\}^{1/2}. \tag{10.22}$$

Above are the inverse functions of the magnetization. After a numerical evaluation of these, then $m(t, b_0)$ can be obtained.

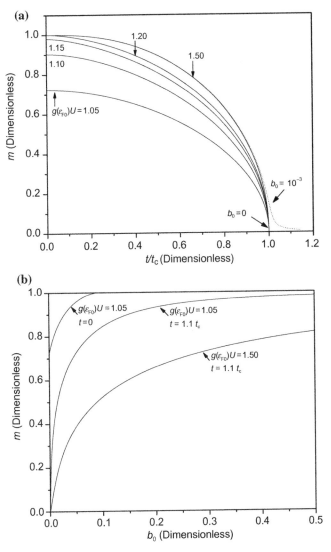

Figure 10.4 Itinerant ferromagnetic reduced magnetization $m = M/N\mu_B$ as a function of (a) reduced temperature t/t_c and (b) reduced external magnetic field $b_0 = \mu_B B_0/\epsilon_{F0}$.

Figure 10.4a presents the reduced magnetization as a function of temperature for both, finite and zero reduced external magnetic field. The case $b_0 = 0$ characterizes the spontaneous magnetization, i.e., $m_{sp} = m(t, b_0 = 0)$. Note $m(0,0)$ decreases by decreasing $g(\epsilon_{F0})U$; and these features will be further analyzed below. For $g(\epsilon_{F0})U < 1$ and m ranging from zero up to one, the inverse equation of state (Eq. (10.22)) returns only complex temperature and therefore we conclude that ferromagnetism ($m_{sp} \neq 0$) only exists for $g(\epsilon_{F0})U > 1$. This important aspect will be considered better below.

Analogously to the localized case, there is a critical temperature t_c, above which the spontaneous magnetization m_{sp} is zero. Also, this quantity will be further analyzed below.

In what concerns the reduced magnetization m as a function of reduced applied magnetic field b_0, it is presented in Figure 10.4b. Above t_c the magnetization behaves as expected: $m(t > t_c, 0) = 0$ and tends to the unity for large values of b_0. In addition, for $t < t_c$ the magnetization starts from m_{sp} and goes to the unity. Note, however, even for values of $g(\epsilon_{F0})U$ in which $m(0,0)$ is smaller than 1, the applied magnetic field leads the magnetization to 1.

10.1.2.2   Critical temperature and the Stoner criterion

We know from the discussion of localized ferromagnetism that the magnetization goes to zero close to the critical temperature and then we need an equation of state for small values of magnetization and $b_0 = 0$. Thus, these conditions into Eq. (10.22) lead to the critical temperature for ferromagnetic itinerant electrons[2]

$$t_c = \frac{2\sqrt{3}}{\pi}[g(\epsilon_{F0})U - 1]^{1/2}. \tag{10.24}$$

The above result is important. For $g(\epsilon_{F0})U < 1$ the critical temperature t_c is complex (nonreal) and therefore the system is paramagnetic. On the other hand, for $g(\epsilon_{F0})U > 1$ the critical temperature t_c is real and positive and therefore the system is ferromagnetic. This fact agrees to what we discussed above. This condition of ferromagnetism is named *Stoner criterion*. Figure 10.5 presents the behavior of t_c as a function of $g(\epsilon_{F0})U$.

There is another way to see the Stoner criterion. Generally speaking, the magnetization is a function of the magnetic field and therefore $M = M(B)$. The inverse function is $B = B(M)$, which can be rewritten as

$$B = \frac{\partial}{\partial M} \int_0^M B(M')dM'. \tag{10.25}$$

Considering the mean field approach (see Eq. (9.37)), the above equation reads as

$$\frac{\partial}{\partial M} \int_0^M B(M')dM' - B_0 - \lambda M = 0. \tag{10.26}$$

However, we know that the equilibrium condition to the free energy is

$$\frac{\partial F}{\partial M} = 0 \tag{10.27}$$

and then a simple comparison of Eqs. (10.27) and (10.26) leads to

$$F = \int_0^M B(M')dM' - B_0 M - \frac{\lambda}{2}M^2. \tag{10.28}$$

[2]Remember, for small values of x
$$(1 + ax)^n \to 1 + nax. \tag{10.23}$$

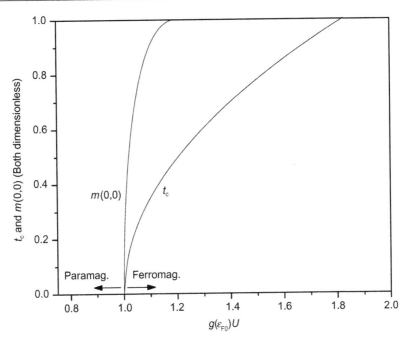

Figure 10.5 Zero Kelvin spontaneous magnetization and critical temperature for itinerant electrons.

Now it is easy to obtain the free energy, since we know the equation of state for small values of magnetization and $t = 0$ (see Eq. (10.18)). Thus,

$$F = \frac{M^2}{2\mu_B^2 g(\epsilon_{F0})} - B_0 M - \frac{\lambda}{2} M^2. \tag{10.29}$$

For the case without external magnetic field ($B_0 = 0$), the free energy resumes as

$$F(B_0 = 0) = \frac{M^2}{2\mu_B^2 g(\epsilon_{F0})}[1 - Ug(\epsilon_{F0})]. \tag{10.30}$$

Note that this last result is the free energy for zero temperature and zero external magnetic field, and therefore, it is the lowest possible magnetic energy. In addition, the pre-factor (outside of [] brackets) is positive. Thus, if $Ug(\epsilon_{F0}) < 1$, then the free energy is positive and the magnetization needs to be zero ($M = 0$), in order to minimize the energy (paramagnetism). On the other hand, if $Ug(\epsilon_{F0}) > 1$, the Gibbs free energy is negative and the magnetization can assume finite values ($M \neq 0$), i.e., the system assumes spontaneous ordering (ferromagnetism). This is the Stoner criterion to itinerant ferromagnetism.

10.1.2.3   Zero Kelvin spontaneous magnetization

Let us now focus our attention to the spontaneous magnetization (defined as $m_{sp} = m(t, b_0 = 0)$), at zero temperature, i.e., $m(0,0)$. To this purpose, Eq. (10.18) is no longer valid, since this one is limited to small values of m, and therefore we must consider Eq. (10.15). Thus,

$$g(\epsilon_{F0})U = \frac{3}{4} \frac{\left[(1 + m(0,0))^{2/3} - (1 - m(0,0))^{2/3}\right]}{m(0,0)}. \tag{10.31}$$

It is easy to see that

$$g(\epsilon_{F0})U \xrightarrow{m(0,0) \to 0} 1 \tag{10.32}$$

and

$$g(\epsilon_{F0})U \xrightarrow{m(0,0) \to 1} \frac{3}{2^{4/3}} \approx 1.19. \tag{10.33}$$

Thus, for $1 \le g(\epsilon_{F0})U \le 1.19$ interval, $m(0,0)$ ranges from 0 up to 1. Figure 10.5 clarifies these ideas.

10.1.2.4   Low temperature magnetic susceptibility

To obtain the magnetic susceptibility for low values of temperature, we need to consider the external magnetic field B_0. Thus, inserting Eq. (10.19) into Eq. (10.18) leads to the mean field magnetization for small values of magnetization and temperature

$$M = \mu_B^2 g(\epsilon_{F0}) \frac{1 - \pi^2 t^2/12}{1 - Ug(\epsilon_{F0})[1 - \pi^2 t^2/12]} B_0. \tag{10.34}$$

The magnetic susceptibility can then be obtained considering the definition of magnetic susceptibility.[3] Thus, for small values of t

$$\chi = \frac{\chi_p}{1 - Ug(\epsilon_{F0})} \left\{ 1 - \frac{\pi^2 t^2}{12[1 - Ug(\epsilon_{F0})]} \right\}, \tag{10.36}$$

where χ_p is the Pauli paramagnetic susceptibility (see Eq. (8.128)). Note that even for those materials that do not satisfy the Stoner criterion to be ferromagnetic (i.e., are paramagnetic), there is a change in the magnetic susceptibility. These materials are in the range $0 \le Ug(\epsilon_{F0}) < 1$ and have an enhancement in the magnetic susceptibility, by a factor $1/[1 - Ug(\epsilon_{F0})]$; and then $\chi > \chi_p$. This is the well-known enhanced paramagnetic susceptibility.

---

[3]Remember

$$\chi = \lim_{H_0 \to 0} \frac{\partial M}{\partial H_0}. \tag{10.35}$$

10.2   Antiferromagnetism

Generally speaking, an antiferromagnetic arrangement has an amount of magnetic moments aligned in opposition to the same amount of magnetic moments of the same kind (see Figure 10.1). As before, we will consider the mean field approach to describe the magnetic properties of an antiferromagnetic arrangement. To this purpose, we need to consider two magnetic sublattices: one with magnetization M_A and magnetic moments interacting between them through a positive mean field parameter λ_{AA}; and other sublattice, with magnetization M_B and the same properties as the sublattice A, however, the magnetic moments interact between them with a positive mean field parameter λ_{BB}.

The above conditions are not enough to obtain an antiferromagnetic arrangement. One magnetic sublattice must be aligned in opposition to the other, through a (negative) mean field parameter $\lambda_{AB} = \lambda_{BA}$. From this consideration, the effective magnetic fields are

$$B_A = B_0 + \lambda_{AA}M_A + \lambda_{AB}M_B \tag{10.37}$$

and

$$B_B = B_0 + \lambda_{BB}M_B + \lambda_{BA}M_A. \tag{10.38}$$

From the above, we must consider

$$\lambda_{AA} = \lambda_{BB} = \alpha\lambda \tag{10.39}$$

and

$$\lambda_{AB} = \lambda_{BA} = -\lambda, \tag{10.40}$$

where $\lambda > 0$.

10.2.1   Magnetization

The magnetization of each sublattice reads as

$$M_A = N_A g j \mu_B B_j \left(\frac{g j \mu_B}{k_B T} [B_0 + \alpha\lambda M_A - \lambda M_B] \right) \tag{10.41}$$

and

$$M_B = N_B g j \mu_B B_j \left(\frac{g j \mu_B}{k_B T} [B_0 + \alpha\lambda M_B - \lambda M_A] \right), \tag{10.42}$$

where $B_j(\cdots)$ is the Brillouin function.

Analogously to the ferromagnetic case, the antiferromagnetic magnetization has no analytical solution and then, to visualize the behavior of this quantity as a function of both, temperature and magnetic field, we need to appeal to a numerical routine. These are presented in Figure 10.6.

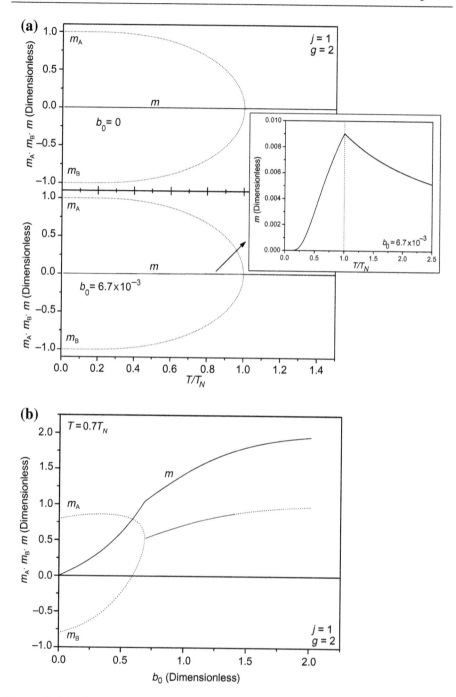

Figure 10.6 Antiferromagnetic reduced magnetization $m_A = M_A/N_A\mu_B$, $m_B = M_B/N_B\mu_B$, and $m = m_A + m_B$ as a function of (a) reduced temperature T/T_N and (b) reduced external magnetic field $b_0 = \mu_B B_0/k_B T_N$. For these curves, we considered $\alpha = 0.01$. Results obtained numerically.

As a function of temperature, for zero applied magnetic field ($b_0 = \mu_B B_0/k_B T_N = 0$), the magnetization of each sublattice behaves like the ferromagnetic case, as expected. Since both sublattices have the same total magnetic moment, thus, the total magnetization $M = M_A + M_B$ is zero above and below the critical temperature T_N, named Néel temperature (see Figure 10.6a). A different scenario arises for a finite applied magnetic field, for which a finite total magnetization appears, since the A sublattice is along B_0 while B sublattice is against. As a consequence, close to T_N this finite total magnetization is bigger, i.e., it peaks at T_N; and then induces a magnetic susceptibility for both, above and below T_N. See Figure 10.6a and its inset.

As a function of magnetic field b_0, this system has an interesting behavior. For temperatures below T_N, each sublattice behaves like a ferromagnetic system, however, the total magnetization contains important features. There is a critical field in which the magnetic energy is equal to the energy to keep the sublattices in opposition and then B sublattice flips from "down" to "up," to be then aligned along the applied magnetic field. However, A and B sublattices interact each other to be antiparallel; and, due to this reason, if B sublattice flips "up," then A sublattice flips "down," to keep the antiferromagnetic character of the system. However, A-B interaction has not enough energy to keep antiparallel and then A sublattice flips back to the original position, i.e., "up." These features are presented in Figure 10.6b.

10.2.2   Néel temperature

Analogously to before, to obtain the critical temperature, named as Néel temperature for the antiferromagnetic case, we need to consider the paramagnetic equation for small values of x, i.e., small values of effective magnetic field. Thus, from Eq. (8.60)

$$M_A = \frac{N_A g^2 \mu_B^2 j(j+1)}{3k_B} \frac{B_A}{T} = \frac{C}{2\mu_0} \frac{B_A}{T}, \tag{10.43}$$

where $N_A = N_B = N/2$ and C is the Curie constant, given by Eq. (10.12). Thus, the above equation reads as

$$M_A = \frac{C}{2\mu_0 T}(B_0 + \alpha \lambda M_A - \lambda M_B) \tag{10.44}$$

and, analogously,

$$M_B = \frac{C}{2\mu_0 T}(B_0 - \lambda M_A + \alpha \lambda M_B). \tag{10.45}$$

The above equations can be rewritten as

$$\begin{pmatrix} u & v \\ v & u \end{pmatrix} \begin{pmatrix} M_A \\ M_B \end{pmatrix} = C B_0 \begin{pmatrix} 1 \\ 1 \end{pmatrix}, \tag{10.46}$$

where

$$u = 2\mu_0 T - \alpha C \lambda \tag{10.47}$$

and

$$v = C\lambda. \tag{10.48}$$

Note Eq. (10.46) defines the magnetization above the critical temperature for this antiferromagnetic system. To determine T_N, we need to consider $B_0 = 0$, as done for the ferromagnetic case. The nontrivial solution for T_N is obtained considering as zero the determinant of the above matrix and thus,

$$T_N = \frac{C\lambda}{\mu_0} \frac{(\alpha + 1)}{2}. \tag{10.49}$$

Above, C is the Curie constant of the whole system, i.e., it depends on $N = N_A + N_B$. Some authors write in Eq. (10.43) only N instead of $N_A = N/2$ and therefore the above result does not have the $1/2$ factor.

10.2.3 Magnetic susceptibility for $T > T_N$

The total magnetization of the system is written as

$$M = M_A + M_B = \frac{C}{2\mu_0 T}(B_A + B_B). \tag{10.50}$$

Considering Eqs. (10.37) and (10.38) and after simple algebraic arrangements we find:

$$M = \frac{C}{T - C\lambda(\alpha - 1)/2\mu_0} H_0. \tag{10.51}$$

Then, the magnetic susceptibility reads as

$$\chi = \frac{C}{T - \theta_p}, \tag{10.52}$$

where

$$\theta_p = \frac{C\lambda}{\mu_0} \frac{(\alpha - 1)}{2}. \tag{10.53}$$

The inverse magnetic susceptibility for this antiferromagnetic system is presented in Figure 10.3. Remember, α rules the intensity of the positive intra-sublattice interaction. If α be smaller than 1 (or even $\alpha \ll 1$), i.e., the ferromagnetic intra-sublattice is weaker than the antiferromagnetic inter-sublattice interaction, then $\theta_p < 0$. In addition, in a different fashion as the ferromagnetic case, this mean field approach leads to $|\theta_p| < T_N$.

10.3 Ferrimagnetism

The ferrimagnetic arrangement has, analogously to the antiferromagnetic case, two magnetic sublattices (see Figure 10.1). However, for the present case, these two have different magnetic character. In other words, these two are, in principle, different ions and therefore have different Lande factor $g_A \neq g_B$ and different total angular momentum $j_A \neq j_B$.

The effective magnetic fields can be written analogously to the antiferromagnetic case

$$B_A = B_0 + \lambda_{AA} M_A + \lambda_{AB} M_B \tag{10.54}$$

and

$$B_B = B_0 + \lambda_{BB} M_B + \lambda_{BA} M_A, \tag{10.55}$$

however, the mean field parameters must be changed as follows

$$\lambda_{AA} = \alpha\lambda, \tag{10.56}$$

$$\lambda_{BB} = \beta\lambda, \tag{10.57}$$

$$\lambda_{AB} = \lambda_{BA} = -\lambda, \tag{10.58}$$

where $\lambda > 0$. Note A species interact each other (via $\alpha\lambda > 0$) in a different fashion than B species (that, on its turn, interacts via $\beta\lambda > 0$). In addition, A sublattice interacts in opposition to B sublattice (via $-\lambda$).

10.3.1 Magnetization

The total magnetization is then obtained by a simple addition of sublattice magnetizations

$$M_A = N_A g_A j_A \mu_B B_{j_A} \left(\frac{g_A j_A \mu_B}{k_B T} [B_0 + \alpha\lambda M_A - \lambda M_B] \right) \tag{10.59}$$

and

$$M_B = N_B g_B j_B \mu_B B_{j_B} \left(\frac{g_B j_B \mu_B}{k_B T} [B_0 + \beta\lambda M_B - \lambda M_A] \right), \tag{10.60}$$

where $B_j(\cdots)$ is the Brillouin function. By construction, each sublattice has its own inner interaction, as well as Lande factor and total angular momentum, and therefore each one behaves like a different ferromagnet; that, on its turn, interacts each other in an antiparallel fashion.

If A sublattice (for instance, $g_A j_A > g_B j_B$), aligned to the external magnetic field B_0, has a (much) stronger inner interaction, i.e., $\alpha \gg \beta$, the total magnetization behaves like a ferromagnetic system (see Figure 10.7a-top). On the other hand, if B sublattice, aligned in opposition to the external magnetic field B_0, has a (much) stronger inner interaction, i.e., $\alpha \ll \beta$, the total magnetization necessarily crosses the temperature axis and turns to be negative. The temperature in which it occurs is named *compensation temperature* and means that $|M_A| = |M_B|$. See Figure 10.7a-bottom for further understanding. Above the critical temperature T_c, the total magnetization has a behavior qualitatively similar to the paramagnetic case. As a function of the external magnetic field (see Figure 10.7b), both, sublattices and total magnetizations behave like the antiferromagnetic case, as presented in Figure 10.6b.

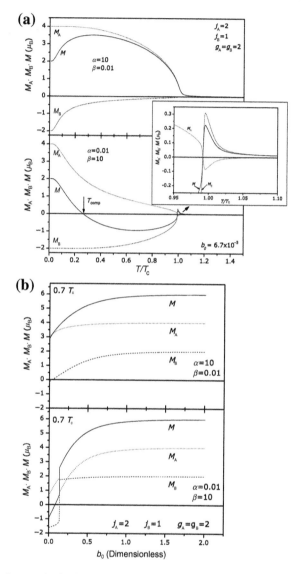

Figure 10.7 Ferrimagnetic absolute magnetization as a function of (a) reduced temperature T/T_c and (b) reduced external magnetic field $b_0 = \mu_B B_0/k_B T_c$.

10.3.2 Critical temperature

As already discussed above, for ferro- and antiferromagnetic systems, to obtain the critical temperature we must consider the equation to describe small values of magnetization, i.e., Eq. (8.60). Thus, it is possible to write

$$M_A = \frac{N_A g_A^2 \mu_B^2 j_A (j_A + 1)}{3 k_B} \frac{B_A}{T} = \frac{C_A}{\mu_0} \frac{B_A}{T} \qquad (10.61)$$

and then

$$M_A = \frac{C_A}{\mu_0 T}(B_0 + \alpha\lambda M_A - \lambda M_B). \tag{10.62}$$

Analogously

$$M_B = \frac{C_B}{\mu_0 T}(B_0 + \beta\lambda M_B - \lambda M_A). \tag{10.63}$$

The above two equations can be rewritten as

$$\begin{pmatrix} u_A & v_A \\ v_B & u_B \end{pmatrix} \begin{pmatrix} M_A \\ M_B \end{pmatrix} = \frac{B_0}{\mu_0 T} \begin{pmatrix} C_A \\ C_B \end{pmatrix}, \tag{10.64}$$

where

$$u_A = 1 - \frac{C_A \alpha \lambda}{\mu_0 T} \quad \text{and} \quad v_A = \frac{C_A \lambda}{\mu_0 T}, \tag{10.65}$$

$$u_B = 1 - \frac{C_B \beta \lambda}{\mu_0 T} \quad \text{and} \quad v_B = \frac{C_B \lambda}{\mu_0 T}. \tag{10.66}$$

To therefore determine the critical temperature below which there is spontaneous ordering, the external magnetic field must be zero: $B_0 = 0$, otherwise the total magnetization is always different of zero and there is no transition. To obtain nontrivial solution, the determinant of the above matrix must be zero and then the critical temperature can be determined:

$$T_c = \frac{\lambda}{2\mu_0} \left\{ (\alpha C_A + \beta C_B) + \left[(\alpha C_A - \beta C_B)^2 + 4C_A C_B \right]^{1/2} \right\}. \tag{10.67}$$

10.3.3 Magnetic susceptibility for T > T_c

The magnetic susceptibility is the response of the magnetization to an external magnetic excitation. Thus, to evaluate this quantity, we need to know the magnetization (for small values of magnetization, since we are interested in the $T > T_c$ range), as a function of B_0. It is not difficult, since the equation we need was written above, in Eq. (10.64). The solution of that equation is easy to be obtained (even by hand) and is reproduced below

$$M_A = \frac{\overline{C}_A T - \overline{C}_A \overline{C}_B \lambda (1 + \beta)}{T^2 - T\lambda(\alpha\overline{C}_A + \beta\overline{C}_B) + \overline{C}_A \overline{C}_B \lambda^2(\alpha\beta - 1)} B_0 \tag{10.68}$$

and

$$M_B = \frac{\overline{C}_B T - \overline{C}_A \overline{C}_B \lambda (1 + \alpha)}{T^2 - T\lambda(\alpha\overline{C}_A + \beta\overline{C}_B) + \overline{C}_A \overline{C}_B \lambda^2(\alpha\beta - 1)} B_0, \tag{10.69}$$

where $\overline{C} = C/\mu_0$. However, we are interested to obtain the magnetic susceptibility and then the total magnetization $M = M_A + M_B$ must be written first (not shown, but easy to be obtained). Then, the magnetic susceptibility reads as

$$\chi = \mu_0 \frac{(\overline{C}_A + \overline{C}_B)T - \overline{C}_A\overline{C}_B\lambda(2 + \alpha + \beta)}{T^2 - T\lambda(\alpha\overline{C}_A + \beta\overline{C}_B) + \overline{C}_A\overline{C}_B\lambda^2(\alpha\beta - 1)}. \tag{10.70}$$

Note the inverse magnetic susceptibility is not linear with temperature, as it was found for both, ferro- and antiferromagnetic systems. It presents a remarkable downturn close to T_c and this feature is one important signature of ferrimagnetic arrangement. However, for high temperature ($T \to \infty$), we can recover the well-known Curie–Weiss law

$$\chi = \frac{C}{T - \theta_p}, \tag{10.71}$$

where

$$C = C_A + C_B \tag{10.72}$$

and

$$\theta_p = \frac{\lambda}{\mu_0} \frac{(\alpha C_A^2 + \beta C_B^2 - 2C_A C_B)}{(C_A + C_B)}. \tag{10.73}$$

See Figure 10.3 for further understanding. Note depending on the values of α and β, the paramagnetic Curie temperature θ_p can assume both, positive and negative values.

Complements

10.A   Experimental examples

These ordered magnetic systems discussed in the present chapter are usual to be found in the literature. As an example, let us consider the manganese oxide series $Pr_{1-x}Ca_xMnO_3$, that has a complex phase diagram, with several magnetic orderings and even phase coexistence of some (further details on these materials can be found in Ref. [7]). Figure 10.8 presents the temperature dependence of both, magnetization under low values of magnetic field ($M/H = \chi$) and inverse magnetic susceptibility ($1/\chi$). Note to consider $M/H = \chi$ the magnetization measurement must be done within the susceptibility range, i.e., under values of magnetic field in which magnetization has a linear dependence on the magnetic field.

Thus, Figure 10.8-top presents a ferromagnetic material ($x = 0.20$), with a magnetic behavior qualitatively the same as the ferromagnetic case described in the present chapter. To obtain the Curie temperature of this material, the inflection point of the $M(T)$ curve was considered, signed with a thick arrow. Note the paramagnetic Curie temperature θ_p is expected to be equal to T_c within the mean field approach; however, for this example, $\theta_p \neq T_c$. It means that the mean field approach is not enough to describe the magnetic properties of this material. Finally, Figure 10.8-bottom presents a ferrimagnetic material ($x = 0.95$). Note the strong downturn on the inverse magnetic susceptibility, typical of ferrimagnetic materials.

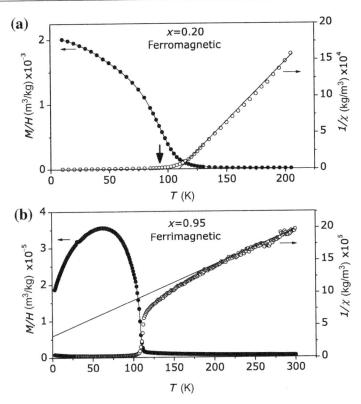

Figure 10.8 Experimental magnetization and inverse susceptibility for some x contents of a manganese oxide series $Pr_{1-x}Ca_xMnO_3$ (further details on these materials can be found in Ref. [7]). Top: $x = 0.20$-ferromagnetic example. The thick arrow represents the Curie temperature, obtained from the inflection point of $M(T)$. Bottom: $x = 0.95$-ferrimagnetic example. Note the strong downturn expected to the inverse magnetic susceptibility of ferrimagnetic materials.

10.B Stoner criterion for some metals

See Figure 10.9. It presents the Stoner parameter U (top panel), density of states at the Fermi level without external magnetic field $g(\epsilon_{F0})$ (middle panel), and $Ug(\epsilon_{F0})$ (bottom panel), for several pure materials (atomic number Z ranging from 3-Lithium up to 49-Indium). Note some transition metals (Iron, Cobalt, and Nickel) satisfy the Stoner criterion due to the increase of the density of states, since the Stoner parameter U is almost linear for this range of atomic numbers Z.

It is interesting to note that some materials, for instance Palladium and bcc-Scandium, have $Ug(\epsilon_{F0})$ close to (but smaller than) 1. These do not satisfy the Stoner criterion and therefore are paramagnetic. However, since $Ug(\epsilon_{F0})$ is close to 1, the paramagnetic susceptibility is indeed enhanced by a $1/[1 - Ug(\epsilon_{F0})]$ factor.

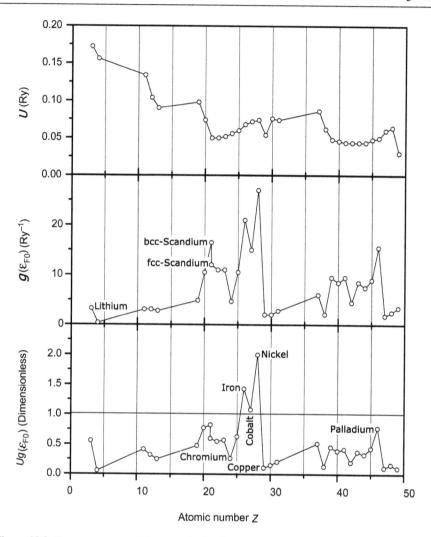

Figure 10.9 Stoner parameter U (top panel), density of states at the Fermi level without external magnetic field $g(\epsilon_{F0})$ (middle panel), and $Ug(\epsilon_{F0})$ (bottom panel), for several pure materials (atomic number Z ranging from 3-Lithium up to 49-Indium) parameter. 1 Ry = 13.6 eV.

10.C   Magnetocaloric Effect

The Magnetocaloric Effect (MCE), discovered in 1881 by Warburg [8], is an exciting property of magnetic materials. This effect can be seen from either an adiabatic or an isothermal process; both due to a change of the applied magnetic field. Considering an adiabatic process, the magnetic material changes its temperature, whereas from an isothermal process, the magnetic material exchanges heat with a thermal reservoir. Figure 10.10 clarifies these processes.

Adiabatic process	Isothermal process

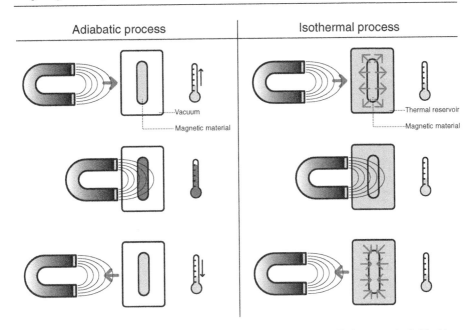

Figure 10.10 Fundamentals of the magnetocaloric effect. An applied magnetic field either changes the temperature of the magnetic material (considering an adiabatic process) or promotes a heat exchange with a thermal reservoir (considering an isothermal process).

From the quantitative point of view, the MCE is measured through the magnetic entropy change $\Delta S = \Delta Q/T$, where ΔQ is the amount of heat exchanged between the thermal reservoir and the magnetic material, when the isothermal process is considered. Analogously, the adiabatic temperature change ΔT characterizes the effect when the adiabatic process is considered. These quantities can also be seen when the magnetic entropy is expressed as a function of temperature for both, with and without applied magnetic field. See Figure 10.11.

It is straightforward the idea to produce a thermo-magnetic cycle based on the isothermal and/or adiabatic processes (like Brayton and Ericsson cycles—see Figure 10.12)); and indeed this idea begun in the late 1920s, when cooling via adiabatic demagnetization was proposed by Debye [9] and Giauque [10]. The process was afterwards demonstrated by Giauque and MacDougall, in 1933, when they reached 250 mK [11]. Since then, the adiabatic demagnetization was used within some contexts; for instance, to cool NASA-XRS detectors (1.5 K) [12]. On the other hand, room temperature magnetic cooling device technology is still in an early phase of development, with no commercially available products and only few prototypes. In August 2001, Astronautics Corporation of America, USA, announced a prototype of room temperature magnetic cooler. This machine has a cooling power of 95 W and uses as the active magnetic material Gd spheres [13]. Later, in March 2003, Chubu Electric and Toshiba, Japan, also announced a room temperature magnetic cooler prototype. This

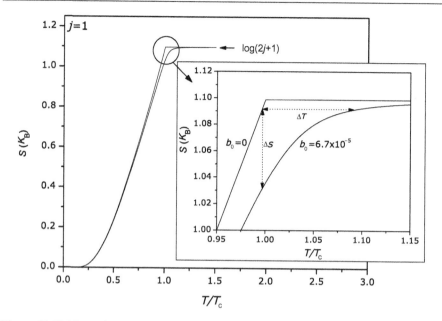

Figure 10.11 Magnetic entropy for a ferromagnetic localized system, obtained numerically considering the mean field approach (see Section 10.1.1 and Eq. (8.70) for further details).

machine has a cooling power of 60 W and uses a layered bed of a Gd-Dy alloy as the active magnetic material [13]. Up to date, ca. 42 prototypes have been published [14].

However, nowadays, the magnetic materials available and studied by the scientific community do not have yet the needed characteristics to be used in large scale, due to technological and/or economic restrictions. For a successful application, we need a material of low cost, nontoxic, good thermal conductivity and with a huge magnetocaloric potential. In this sense, most of the research developed worldwide is devoted to explore and optimize the magnetocaloric properties of known materials, as well as to seek for new magnetocaloric features in new materials.

In addition to the spin entropy, other mechanisms sensible to the magnetic field can be useful to maximize the magnetic entropy change, namely lattice, charge, and orbital entropies; since these entities, in certain cases, can also be ruled by the applied magnetic field. In this sense, some materials with strong electronic correlation have coupled magneto-structural transitions, maximizing the magnetocaloric potential.

10.C.1   Ideal thermodynamic processes

The Magnetocaloric Effect can be understood, as mentioned above, via either isothermal or adiabatic processes; and these processes are characterized by the magnetic entropy change and the adiabatic temperature change, respectively. Below, we describe how these quantities are measured.

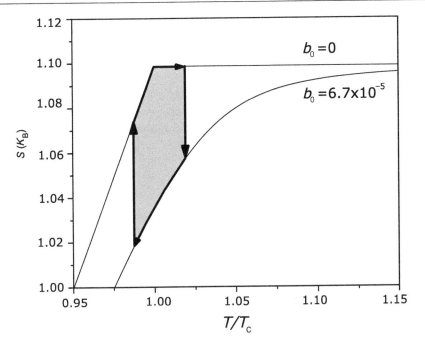

Figure 10.12 Ericsson cycle for the magnetocaloric effect.

Magnetic entropy change

Section 4.4 described how to obtain Maxwell relations and, for the purpose of the present study, we need one of them, more precisely that in Eq. (4.43), reproduced below

$$\left.\frac{\partial M}{\partial T}\right|_B = \left.\frac{\partial S}{\partial B}\right|_T . \tag{10.C.1}$$

After integration, the above equation reads as:

$$\Delta S(T, \Delta B) = \int_{B_i}^{B_f} \left.\frac{\partial M}{\partial T}\right|_B dB . \tag{10.C.2}$$

Thus, to obtain the magnetic entropy change we need to measure magnetization as a function of magnetic field and temperature, i.e., $M(B, T)$.

There is another way to obtain the magnetic entropy change: from specific heat, defined in Eq. (4.47)

$$C_B = T \left.\frac{\partial S}{\partial T}\right|_B . \tag{10.C.3}$$

Analogously to before, a simple integration leads to

$$\Delta S(T, \Delta B) = \int_0^T \frac{C_B - C_0}{T} dT,$$ (10.C.4)

where $C_0 = C_B(B = 0)$. Figure 10.13a presents the behavior of the magnetic entropy change for a localized angular momentum, obtained numerically considering the mean field approach (see Section 10.1.1 and Eq. (8.70) for further details). Note this quantity peaks at T_c. Figure 10.13b presents the experimental magnetic entropy change for a simple ferromagnetic material: $PrNi_2Co_3$. Further information on the material can be found in [15].

Adiabatic temperature change

Entropy is defined as a function of temperature T and magnetic field B

$$S = S(T, B)$$ (10.C.5)

and then

$$dS = \left.\frac{\partial S}{\partial T}\right|_B dT + \left.\frac{\partial S}{\partial B}\right|_T dB.$$ (10.C.6)

The adiabatic condition means $dS = 0$ and therefore, considering the definition of specific heat on Eq. (10.C.3) and the Maxwell relation on Eq. (10.C.1), then Eq. (10.C.6) reads as

$$dT = -\frac{T}{C_B} \left.\frac{\partial M}{\partial T}\right|_B dB.$$ (10.C.7)

Finally, the adiabatic temperature change is

$$\Delta T(T, \Delta B) = -\int_{B_i}^{B_f} \frac{T}{C_B} \left.\frac{\partial M}{\partial T}\right|_B dB.$$ (10.C.8)

Thus, in conclusion, to obtain the adiabatic temperature change, we need to measure $M(B, T)$ and $C(B, T)$. However, there is another way to measure this quantity, i.e., directly applying a magnetic field to the sample (adiabatically) and measuring the corresponding temperature change.

10.C.2   Materials

Below, we describe families of materials promising for applications in magnetic cooling devices to work around room temperature (RT):

R-G family (R: rare earth, G: metalloid): Since the discovery of the huge MCE in $Gd_5Si_2Ge_2$ (35 J/kgK@280 K), in 1997 [16], hundreds of papers have been published. That value is due to a coupled magneto-structural first-order transition, driven by temperature [17], magnetic field [18], and pressure [19]. Also of interest is the

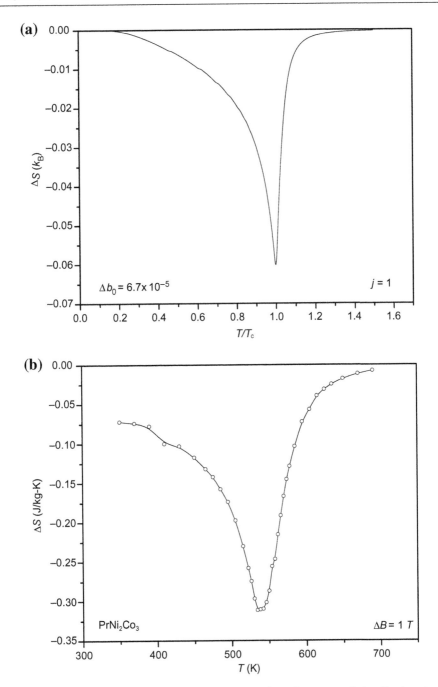

Figure 10.13 (a) Theoretical magnetic entropy change for a ferromagnetic localized system, obtained numerically considering the mean field approximation (see Section 10.1.1 and Eq. (8.70) for further details). (b) Experimental magnetic entropy change for a simple ferromagnetic material: $PrNi_2Co_3$. Further details on the material in Ref. [15].

work of Morellon et al. [19] that could merge the structural and magnetic transition of $Tb_5Si_2Ge_2$ compounds by applying 8.6 kbar of hydrostatic pressure, increasing therefore the MCE.

R-M-G family (M: transition metal): Another series with magneto-structural coupling is $La(Fe_xSi_{1-x})_{13}$ [20]. For instance, MCE for $x = 0.88$ reaches 26 J/kgK @ 188 K. These compounds under a negative pressure (expansion of the unit cell), via insertion of Hydrogen, shift the transition temperature up to RT [21]; while positive pressure (compression of the unit cell), via hydrostatic pressure, increases the MCE value, decreasing, however, the transition temperature [22].

Mn-M-G family: MnAs is a compound that presents a coupled magneto-structural transition of first order, resulting therefore in a huge MCE: 32 J/kgK @ 318 K [23]. Gama et al. [24] applied 2.2 kbar in MnAs compound and increased the MCE up to 267 J/kgK @ 280 K.

Concerning the Heusler alloys Ni_2MnGa, this material has a decoupled magneto-structural transition of about 100 K. Hu et al. [25] were the first to report the magnetocaloric properties of this compound; and, after, Zhou et al. [26] reported a coupled magneto-structural transition for the out of stoichiometry $Ni_{55}Mn_{19}Ga_{26}$, with a huge, but narrow, MCE: 20 J/kgK @ 317 K.

Manganites family ($RMnO_3$): The MCE of mixed-valency manganites was first measured by Morelli et al. in 1996, for thick films of $La_{2/3}(Ca/Sr/Ba)_{1/3}MnO_3$ [27]. The magneto-structural coupling [28] and the numerous possibilities of exchanging elements in synthesis make manganites a rich field in MCE study. The substitution of La by other rare-earth ion in $La_{2/3}Ca_{1/3}MnO_3$ manganite can tune the transition temperature and enhance magnetocaloric properties [29]. Some manganites systems show also a magneto-structural coupling [7].

Intermetallics family (R-M): There are several subfamilies, namely the Laves Phase compounds, as RCo_2[30], RAl_2 [31], and RNi_2[32], but all of these are not suitable for applications around room temperature due to their low values of MCE. On the contrary, those systems are the reason for several theoretical models, due to their interesting magnetocaloric properties (from academic point of view) [32]. Other compounds can also be cited: Nd_2Fe_{17} (5.9 J/kgK @ 325 K) [33], Gd_7Pd_3 (6.5 J/kgK @ 323 K) [34], and Gd_4Bi_3 (2.7 J/kgK @ 332 K) [35].

The above values of MCE correspond to 5 T of magnetic field change. Those families above described, and their corresponding compounds, have problems that avoid their immediate application in a magnetic cooling device: (i) first-order magnetic transition and a consequent thermal hysteresis, producing then energy losses during the thermo-magnetic cycle; (ii) high pure rare-earth metals, making them economically unviable; (iii) elements that need special handling, for instance the poisonous arsenic; (iv) narrow magnetic entropy change curves, avoiding therefore a wide thermo-magnetic cycle; and, finally, (v) low values of MCE, decreasing the cooling power of the device.

10.D   Undesired temperature independent susceptibility

It is common to find materials with a "parasitic" temperature independent contribution to the susceptibility, due to either diamagnetism (negative contribution), or (Pauli) paramagnetism (positive contribution). Both cases change the linear behavior of the inverse magnetic susceptibility and then can be considered as undesired to the purpose of analysis of the paramagnetic effective moment and paramagnetic Curie temperature.

The magnetic susceptibility at high temperature is given by

$$\chi = \frac{C}{T - \theta_p} + \chi_0. \tag{10.D.1}$$

This result comes from Eq. (10.11), where we added a temperature independent contribution χ_0. From the above equation it is not possible to obtain the fundamental quantities C and θ_p. Note from Figure 10.14 there is a loss of linearity of the inverse susceptibility due to the temperature independent contribution. A solution is to proceed the first derivative of the magnetic susceptibility with respect to the temperature, since this procedure eliminates this undesired temperature independent contribution. Thus,

$$\frac{d\chi}{dT} = -\frac{C}{(T - \theta_p)^2} \tag{10.D.2}$$

and

$$\left|\frac{d\chi}{dT}\right|^{-1/2} = \frac{1}{\sqrt{C}}T - \frac{\theta_p}{\sqrt{C}}. \tag{10.D.3}$$

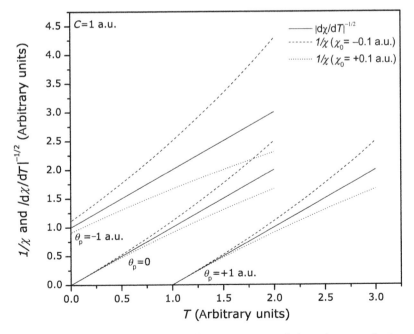

Figure 10.14 Inverse susceptibility $1/\chi$ with a temperature independent contribution (see Eq. (10.D.1) and dotted and dashed lines in this figure). Note χ_0 introduces a nonlinear behavior into the Curie–Weiss law; and this problem can be overcome considering $|d\chi/dT|^{-1/2}$ (see Eq. (10.D.3) and the solid lines in this figure).

In other words, if the material has a significant temperature independent contribution, it is possible to either under- or overestimate important quantities like C and θ_p when performing $1/\chi$ plots. Thus, the easier way is to deal with $|d\chi/dT|^{-1/2}$ instead.

To understand better these ideas, let us consider the KNaCuSi$_4$O$_{10}$ compound. Note only the copper ion has unpaired electrons (then, it has a finite spin magnetic moment), while the other ions only contribute diamagnetically to total magnetic moment of the species. Thus, since there are 16 diamagnetic ions to only one ion with spin momentum ($s = 1/2$), a diamagnetic contribution comparable to the magnetic moment of the copper ions is expected, and indeed it occurs. See Ref. [36] for further details.

Figure 10.15-top presents the inverse magnetic susceptibility of the compound. Note there is a deviation from the Curie law around 100 K, due to the (undesired) diamagnetic

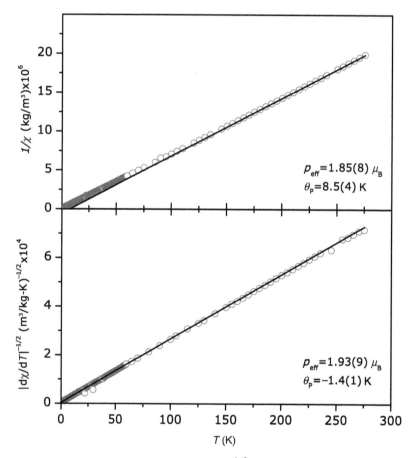

Figure 10.15 Comparison of $1/\chi$ (top) and $|d\chi/dT|^{-1/2}$ (bottom) for KNaCuSi$_4$O$_{10}$ compound [36], that has a diamagnetic temperature independent contribution to the magnetic susceptibility χ. Note it is a mistake to obtain the paramagnetic effective moment p_{eff} and paramagnetic Curie temperature θ_p from $1/\chi$ when there is a temperature independent contribution; and the correct way is to consider $|d\chi/dT|^{-1/2}$.

contribution, that induces to a mistake obtaining the paramagnetic effective moment p_{eff} and paramagnetic Curie temperature θ_p. Figure 10.15-bottom presents the quantity suggested in this subsection: $|d\chi/dT|^{-1/2}$. As an example of discrepancy between these two approaches, note the obtained p_{eff} and θ_p, presented in Figure 10.15. For the first approach, $\theta_p < 0$ and then predicts an antiparallel alignment between copper ions; while from the second approach, $\theta_p > 0$ and then predicts a parallel alignment of those. The conclusion from the first approach (i.e., consider the inverse susceptibility for systems that has a significant temperature independent contribution to the magnetic susceptibility) is a mistake and therefore the second approach is recommended.

11 Landau Theory

11.1 Fundamentals

Landau theory was developed by Lev Davidovich Landau (1908–1968), a Russian physicist. This model is based on two basic assumptions: the free energy F of the system must be analytical and follows the symmetry of the Hamiltonian. In this sense, the free energy is considered as a function of powers of an order parameter ξ and reads as

$$F(\xi, T) = -g_1\xi + \frac{1}{2}g_2\xi^2 + \frac{1}{4}g_4\xi^4 + \frac{1}{6}g_6\xi^6 + \cdots , \tag{11.1}$$

where the multiplier parameters, g_i, can be, in principle, dependent on the temperature: $g_i = g_i(T)$. Note that we added to the free energy an extra term, linear on ξ, that represents an external force that breaks the symmetry of the system. It is negative (and g_1 positive) to favor the minimum of energy at positive values of the order parameter.

11.2 Second-order phase transition

This case considers that the order parameter must be zero for temperatures equal and above a certain critical temperature T_c and different of zero (finite value), below this critical temperature. The case in which all of the g_i parameters are positive is not reasonable, because there is only one solution to the problem: $\xi = 0$. Thus, at least one parameter must be negative and, in addition, the parameter of the highest power on ξ must be positive, to ensure that F is increasing for ξ approaching the unity (this requirement ensures a stable minimum of the potential for finite values of ξ). Considering the simplest problem, let us make $g_1 = 0$ (i.e., there is no external force), and only powers up to ξ^4. Then, we have two parameters to rule the phase transition of this system. We can consider

$$g_2 = \alpha(T - T_0),$$
$$g_4 > 0,$$
$$g_6 = 0 \tag{11.2}$$

Fundamentals of Magnetism. http://dx.doi.org/10.1016/B978-0-12-405545-2.00011-4

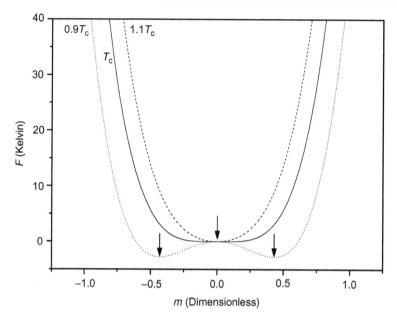

Figure 11.1 Free energy as a function of the order parameter. This figure deals with the case of a classical magnetic moment and therefore the order parameter is the magnetization. There are three cases: $T < T_c$, $T = T_c$, and $T > T_c$, without external force g_1. Note $T_c = T_0$—see Eq. (11.11).

and then, for temperatures above T_0, both, g_4 and g_2 are positives; and the solution is $\xi = 0$. Below T_0, only g_2 is negative and the minimum at $\xi = 0$ turns to be a maximum. As a consequence, a local minimum on the free energy F appears for $\xi \neq 0$. Figure 11.1 clarifies these words.

The free energy can then be written as

$$F(\xi, T) = \frac{1}{2}\alpha(T - T_0)\xi^2 + \frac{1}{4}g_4\xi^4. \tag{11.3}$$

To find a minimum we need to satisfy the condition

$$\frac{\partial}{\partial \xi} F(\xi, T) = 0 = \alpha(T - T_0)\xi + g_4\xi^3 \tag{11.4}$$

and therefore there are three possible solutions

$$\xi = 0 \quad \text{and} \quad \xi_{\pm} = \pm\left(\frac{\alpha(T_0 - T)}{g_4}\right)^{1/2}. \tag{11.5}$$

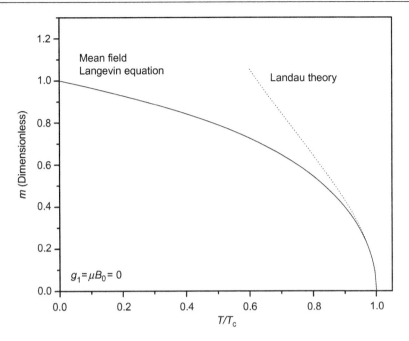

Figure 11.2 Magnetization (order parameter), as a function of temperature for both, mean field Langevin equation (obtained via numerical calculation) and Landau theory.

Above, we presented the behavior of the order parameter as a function of temperature. For $T < T_0$, there are two (equivalent) minima and the solutions are $\xi_{\pm}$. At $T \geq T_0$ there is only one real solution ($\xi = 0$). Note that, for this case, $T_c = T_0$. Figure 11.2 clarifies these words. It is important to stress that Landau theory is only valid for small values of the order parameters.

Now, let us consider the external force $(-g_1\xi)$ and add it to Eq. (11.3)

$$F(\xi, T) = -g_1\xi + \frac{1}{2}\alpha(T - T_0)\xi^2 + \frac{1}{4}g_4\xi^4. \tag{11.6}$$

The equilibrium condition to the above case is

$$\frac{\partial}{\partial\xi}F(\xi, T) = 0 = -g_1 + \alpha(T - T_0)\xi + g_4\xi^3. \tag{11.7}$$

We still have three solutions, but now it depends on g_1. One to consider is

$$\xi = -\left(\frac{2}{3}\right)^{1/3}\frac{g_2}{p} + \left(\frac{1}{2}\right)^{1/3}\left(\frac{1}{3}\right)^{2/3}\frac{p}{g_4}, \tag{11.8}$$

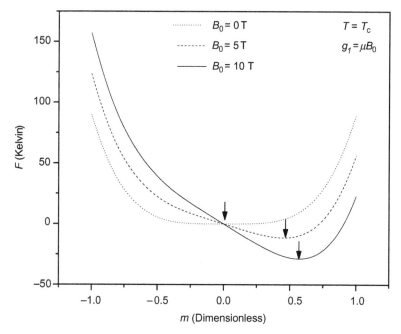

Figure 11.3 Free energy as a function of the order parameter. This figure deals with the case of a classical magnetic moment and therefore the order parameter is the magnetization. For this case, $T = T_c$ and there is an external force g_1.

where

$$p = \left(9g_1g_4^2 + \sqrt{3}\sqrt{4g_2^3g_4^3 + 27g_1^2g_4^4}\right)^{1/3}. \tag{11.9}$$

It is easy to verify that

$$g_1 = 0 \Rightarrow \xi = 0. \tag{11.10}$$

Thus, the above solution (Eq. (11.8)) recovers $\xi = 0$ for the case without external force and thus we are dealing with the evolution of $\xi = 0$ due to g_1 influence (the other two solutions do not interest, as discussed further below).

In addition,

$$T = T_0 = T_c \Rightarrow g_2 = 0 \Rightarrow \boxed{\xi(T_c) = \left(\frac{g_1}{g_4}\right)^{1/3}} \tag{11.11}$$

Note the above solution is real. Figure 11.3 presents the free energy as a function of the order parameter (for this case, the magnetization—further details in the next subsection), at $T = T_c$ for some values of external force. Note there is only one real solution (among the three possible), and the other two are on the complex plane and then do not interest (for $T \geq T_c$). To complete our analysis, we can consider that below T_c (with $g_1 \neq 0$), the order parameter is high enough for the Landau theory

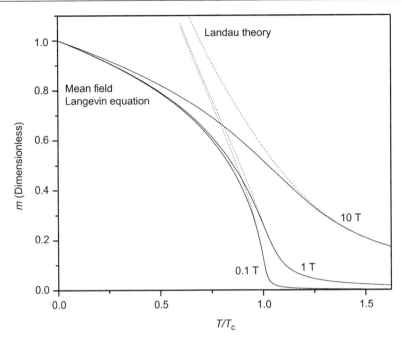

Figure 11.4 Magnetization (order parameter), as a function of temperature for both, mean field Langevin equation (obtained via numerical calculation) and Landau theory. For this case, there is an external force (magnetic field, for the example discussed).

to be valid and then we do not need to know the other solutions. Finally, Figure 11.4 presents the order parameter as a function of the temperature, for some values of external force g_1.

Let us now understand how the susceptibility of the order parameter due to the external force is. It is defined as

$$\chi = \lim_{g_1 \to 0} \frac{\partial \xi}{\partial g_1}. \tag{11.12}$$

To get this susceptibility we must derive both sides of Eq. (11.7) with respect to g_1 and then, to the result, consider the $g_1 \to 0$ limit. These steps are of easy evaluation and we obtain

$$\chi = \frac{1}{g_2 + 3\xi_0^2 g_4}, \tag{11.13}$$

where

$$\xi_0 = \lim_{g_1 \to 0} \xi. \tag{11.14}$$

11.2.1 Connections with a classical spin system

Let us now consider a real case to better understand the Landau theory: a classical spin system. For this purpose, let us go back some chapter, and recall Eq. (8.46), that describes the behavior of this system as a function of temperature and magnetic field. Thus

$$M(x) = \mu \left[\coth (x) - \frac{1}{x} \right],$$
(11.15)

where μ is the saturation value of the magnetization in μ_B units and

$$x = \frac{\mu B}{k_B T}.$$
(11.16)

However, within the mean field approximation, it is possible to write

$$x = \frac{\mu}{k_B T}(B_0 + \lambda M).$$
(11.17)

For further details on the mean field approximation, see Chapter 9.

The development shown below is a little trick. Considering

$$x(M) = M^{-1}(M)$$
(11.18)
$$= \frac{\partial}{\partial M} \int_0^M M^{-1}(M')dM',$$

then Eq. (11.17) turns to be

$$\frac{k_B T}{\mu} \frac{\partial}{\partial M} \int_0^M M^{-1}(M')dM' - B_0 - \lambda M = 0.$$
(11.19)

However, zero is also the first derivative of the free energy as a function of the order parameter, i.e., the equilibrium condition of the system

$$\frac{\partial F}{\partial M} = 0$$
(11.20)

and then, after a simple integration we obtain

$$F = \frac{k_B T}{\mu} \int_0^M M^{-1}(M')dM' - B_0 M - \frac{\lambda}{2} M^2.$$
(11.21)

For small values of x, the equation of state $M(x)$ is

$$M(x) \xrightarrow{x \to 0} \frac{\mu}{3}x - \frac{\mu}{45}x^3 + \frac{2\mu}{945}x^5$$
(11.22)

and the inverse series[1]

$$x(M) = M^{-1}(M) = 3\left(\frac{M}{\mu}\right) + \frac{9}{5}\left(\frac{M}{\mu}\right)^3 + \frac{297}{175}\left(\frac{M}{\mu}\right)^5. \qquad (11.26)$$

Now the integral above can be easily evaluated

$$\int_0^M M^{-1}(M')dM' = \frac{3}{2}\frac{M^2}{\mu} + \frac{9}{20}\frac{M^4}{\mu^3} + \frac{297}{1050}\frac{M^6}{\mu^5} \qquad (11.27)$$

and the free energy assumes a simple form, analogous to that proposed by Landau theory and described in the previous section (Eq. (11.1))

$$F = -(B_0\mu)m + \left(\frac{3}{2}k_BT - \frac{1}{2}\lambda\mu^2\right)m^2 + \left(\frac{9}{20}k_BT\right)m^4 + \left(\frac{297}{1050}k_BT\right)m^6, \qquad (11.28)$$

where $m = M/\mu$ (dimensionless). A simple inspection of both free energies (Eqs. (11.1) and (11.28)), leads us to obtain

$$g_1 = B_0\mu, \qquad (11.29)$$
$$g_2 = \alpha(T - T_0),$$
$$g_4 = \frac{9}{5}k_BT,$$

where

$$\alpha = 3k_B \quad \text{and} \quad T_0 = T_c = \frac{\lambda\mu^2}{3k_B}. \qquad (11.30)$$

---

[1] Considering an expansion

$$y = a_1x + a_2x^2 + a_3x^3 + \cdots, \qquad (11.23)$$

its inverse is

$$x = A_1y + A_2y^2 + A_3y^3 + \cdots, \qquad (11.24)$$

where the parameters on capital letters are:

$$A_1 = a_1^{-1}, \qquad (11.25)$$
$$A_2 = -a_1^{-3}a_2,$$
$$A_3 = a_1^{-5}(2a_2^2 - a_1a_3),$$
$$A_4 = a_1^{-7}(5a_1a_2a_3 - a_1^2a_4 - 5a_2^3),$$
$$A_5 = a_1^{-9}(6a_1^2a_2a_4 + 3a_1^2a_3^2 + 14a_2^4 - a_1^3a_5 - 21a_1a_2^2a_3).$$

$\cdots$

Note g_4 is positive; it ensures that F increases when m tends to the unity and then it has a stable minimum at finite values of the magnetization. Thus, for the second-order transition, expansion of the free energy up to the fourth order is enough. This behavior is presented in Figure 11.1.

From Eq. (11.5), we can obtain the evolution of the magnetization as a function of the temperature without the external force (i.e., the magnetic field B_0)

$$m = \sqrt{\frac{5}{3}} \left(\frac{T_c}{T} - 1 \right)^{1/2} \tag{11.31}$$

considering only the positive solution. This result and that obtained via mean field approach (see Chapter 9 for further details) are presented in Figure 11.2. Note the agreement between these two models for $m \lesssim 1/5$.

From the above equation, another important result arises: the critical exponent. Note that the magnetization is scaled with the temperature by means of an exponent $\beta = 1/2$.

Let us now turn on the external force, that, for this example, is the external magnetic field. The magnetization is now given by Eq. (11.8) and the g_i parameters of the free energy are still given by Eq. (11.29). Figure 11.4 presents the mean field Langevin equation with an external applied field (see Eqs. (11.15) and (11.17)), and Eq. (11.8).

From Eq. (11.11), it is also possible to obtain the magnetization as a function of magnetic field at $T = T_c$

$$m = \left(\frac{5}{9} \frac{\mu}{k_B T_c} \right)^{1/3} (B_0)^{1/3}. \tag{11.32}$$

Considering that the magnetization at this critical temperature follows the critical law: $m \propto B_0^{1/\delta}$, then we can obtain the critical exponent $\delta = 3$. It can be seen in Figure 11.5.

The magnetic susceptibility can also be obtained from Eq. (11.13), however, it needs to be rewritten, based on the definition of magnetic susceptibility

$$\chi_m = \lim_{H_0 \to 0} \frac{\partial M}{\partial H_0} = \mu_0 \mu \lim_{B_0 \to 0} \frac{\partial m}{\partial B_0} = \mu_0 \mu \left(\lim_{g_1 \to 0} \frac{\partial \xi}{\partial g_1} \right) \frac{\partial g_1}{\partial B_0} = \mu_0 \mu^2 \chi, \tag{11.33}$$

where χ was defined above (Eq. (11.13)). To go further, we need to know ξ_0 (given by Eq. (11.14)). It is the magnetization without magnetic field, i.e., $m_0(T \geq T_c) = 0$ and $m_0(T < T_c)$ are given by Eq. (11.31). Then

$$\chi_m = \begin{cases} C(T - T_c)^{-1}, & T \geq T_c, \\ \frac{C}{2}(T_c - T)^{-1}, & T < T_c, \end{cases} \tag{11.34}$$

where

$$C = \frac{\mu_0 \mu^2}{3k_B}. \tag{11.35}$$

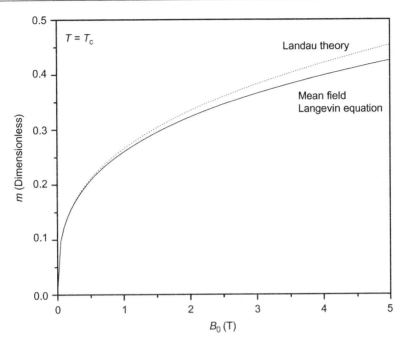

Figure 11.5 Magnetization (order parameter), as a function of magnetic field for both, mean field Langevin equation (obtained via numerical calculation), and Landau theory.

Finally, the magnetic susceptibility is expected to depend on the temperature, around the critical temperature, as $\chi \propto |T_c - T|^{-\gamma}$; and then $\gamma = 1$.

11.3 First-order phase transition

The procedure to obtain first-order transition is similar to the previous one; however, now, we must consider

$$g_2 = \alpha(T - T_0), \tag{11.36}$$
$$g_4 < 0,$$
$$g_6 > 0.$$

Thus, we can write

$$F(\xi, T) = -g_1\xi + \frac{1}{2}g_2\xi^2 - \frac{1}{4}|g_4|\xi^4 + \frac{1}{6}g_6\xi^6. \tag{11.37}$$

Let us consider first the case without external force, i.e., $g_1 = 0$. Then, we must apply the equilibrium condition

$$\frac{\partial}{\partial \xi}F(\xi, T) = 0 = g_2\xi - |g_4|\xi^3 + g_6\xi^5. \tag{11.38}$$

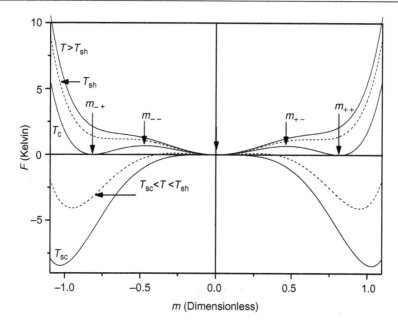

Figure 11.6 Free energy as a function of the order parameter. This figure deals with the case of a classical magnetic moment and therefore the order parameter is the magnetization. There are five temperatures: $T > T_{sh}, T_{sh}, T_c, T_{sc} < T < T_{sh}$, and T_{sc}, without external force g_1. The arrows mean the possible equations of states (see Eq. (11.40)).

Note above we already considered the minus sign of the g_4 parameter and from now on, for the sake of simplicity, we will write: $|g_4| = g_4$. The above polynomial goes up to ξ^5 and, obviously, has five roots. First, the trivial one

$$\xi = 0 \tag{11.39}$$

and then, considering $x = \xi^2$, we obtain the other four roots

$$\xi^2 = x_\pm = \frac{g_4 \pm \sqrt{g_4^2 - 4g_2g_6}}{2g_6}. \tag{11.40}$$

Two solutions at $\xi_- = -\sqrt{x_\pm}$ are not interesting to our purpose, since these are symmetric to $\xi_+ = +\sqrt{x_\pm}$. In this sense, let us focus our attention to $\xi_{++} = +\sqrt{x_+}$ and $\xi_{+-} = +\sqrt{x_-}$, that correspond, respectively, and for $g_2 > 0$, to a minimum and a maximum at $\xi \neq 0$. See Figure 11.6 to clarify these words.

For $g_2 > 0$, the free energy close to $\xi = 0$ is positive; however, g_4 is negative and then the free energy must change its concavity (here a maximum appears, at ξ_{+-}). Finally, g_6 is positive and the free energy must change its concavity again; and here is the local minimum, at ξ_{++}.

The critical temperature T_c is defined when the free energy of the local minimum at $\xi = 0$ is the same that at ξ_{++}. There is only on possible solution

$$F(0, T_c) = F(\xi_{++}, T_c) = 0 \tag{11.41}$$

and then, for $T = T_c$ the relation below is true

$$g_2 = \frac{3}{16} \frac{g_4^2}{g_6}. \tag{11.42}$$

See Figure 11.6 for further understanding.

This model is quite rich and the free energy can change its maxima/minima; and even lost those to the complex plane. In this direction, the maximum at ξ_{+-} goes to zero when the free energy close to $\xi = 0$ changes from positive to negative. Intuitively, it occurs at $g_2 = 0$, i.e., at $T = T_0$; and the proof of this result is left as an exercise. See Figure 11.6. Generally speaking, we can rename the temperature below which there is only one real solution at ξ_{++} as supercooling T_{sc}. Note for the present case, without external force g_1, $T_{sc} = T_0$.

Finally, the local minimum at ξ_{++} disappears, i.e., turns to be an inflection point at

$$g_4^2 - 4g_2 g_6 = 0. \tag{11.43}$$

Well, this condition is the threshold below which the real solution at this point goes to the complex plane. The temperature of this condition can be named as superheating T_{sh} and satisfies

$$g_2 = \frac{1}{4} \frac{g_4^2}{g_6}. \tag{11.44}$$

See Figure 11.6.

Now, it is important to understand why these evaluations. Between T_{sc} and T_{sh} the free energy has two minima. Below T_{sc} only one minimum at ξ_{++}, and above T_{sh} also only one minimum, however, at $\xi = 0$.

Consider this system at high temperature (above T_{sh}); then, it is at $\xi = 0$. Cooling down, the system passes through T_c without changes and then, only at T_{sc}, when the energy barrier between the local minimum at $\xi = 0$ and the one at ξ_{++} disappears, it goes suddenly from $\xi = 0$ to ξ_{++}. Note there is a discontinuity on the order parameter as a function of temperature. Now, let us warm the system from this low temperature (below T_{sc}), at the ξ_{++} solution. At T_c, again, the system passes without changes and only at T_{sh} the solution goes from ξ_{++} to $\xi = 0$ suddenly. This process leads to a thermal hysteresis, measured as $\Delta T = T_{sh} - T_{sc}$. This hysteresis can be expressed as (considering Eq. (11.36))

$$\Delta T = \frac{g_2(T_{sh})}{\alpha}. \tag{11.45}$$

Figure 11.7 clarifies these ideas, where we can observe the hysteresis, normally observed in first order phase transitions. Note this figure justifies the name of these critical temperatures: supercooling and superheating.

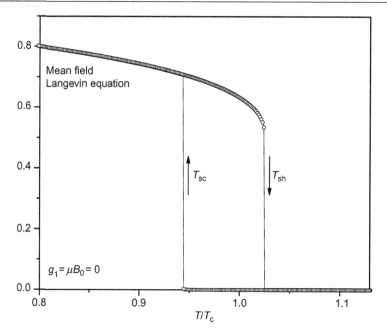

Figure 11.7 Magnetization (order parameter), as a function of temperature. Here, magnetization is given by the mean field Langevin equation (obtained via numerical calculation), considering the M^3 term to the mean field approximation.

Let us now turn on the external force g_1. The problem thus increases in difficulty, and the transition temperatures (T_{sh} and T_{sc}), as well as the order parameter, are not simple to be obtained.

The equilibrium condition, analogously to Eq. (11.38), reads as

$$\frac{\partial}{\partial \xi} F(\xi, T) = 0 = -g_1 + g_2 \xi - g_4 \xi^3 + g_6 \xi^5. \tag{11.46}$$

To obtain some information of the system, we can use regular perturbation theory and consider the order parameter as a function of powers of the external force

$$\xi = x_0 + g_1 x_1 + g_1^2 x_2 + \mathcal{O}(g_1^3) \tag{11.47}$$

and then leads the above equation into Eq. (11.46). After that, we need to make zero the terms of $\mathcal{O}(g_1^n)$. For instance, for the terms without g_1, we must consider

$$\mathcal{O}(g_1^0) = 0 \tag{11.48}$$

and then we recover Eq. (11.38)

$$g_2 x_0 - g_4 x_0^3 + g_6 x_0^5 = 0, \tag{11.49}$$

in which the solutions were already discussed in this section.

Analogously to above, we do the same for the higher order terms of g_1 and obtain x_1 and x_2

$$\mathcal{O}(g_1^1) = 0 \ \Rightarrow \ x_1 = (g_2 - 3g_4 x_0^2 + 5g_6 x_0^4)^{-1}, \tag{11.50}$$
$$\mathcal{O}(g_1^2) = 0 \ \Rightarrow \ x_2 = x_0 x_1^3 (3g_4 - 10g_6 x_0^2). \tag{11.51}$$

The simplest result we can derive from this case with external force is the deviation of the previous solution at $\xi = 0$ due to g_1. Thus, considering $x_0 = 0$ we obtain

$$x_0 = 0 \ \Rightarrow \ x_1 = g_2^{-1} \ \Rightarrow \ x_2 = 0 \tag{11.52}$$

and then, placing the above result into Eq. (11.47) the order parameter reads as

$$\xi = \frac{g_1}{g_2}. \tag{11.53}$$

Obviously, T_{sc} and T_{sh} depend on g_1, but it is not a trivial task to be evaluated here.

11.3.1   Connections with a classical spin system

As done before, it is possible to connect the previous results with a classical spin system. To go further we first need to consider a M^3 correction to the mean field approximation. Thus, Eq. (11.17) turns to be

$$x = \frac{\mu}{k_B T} (B_0 + \lambda M + \lambda' M^3), \tag{11.54}$$

where λ' is generally smaller than λ (say, $\lambda = 10^2 \lambda'$). Following the same procedure of Section 11.2.1, Eq. (11.21) changes to

$$F = \frac{k_B T}{\mu} \int_0^M M^{-1}(M') dM' - B_0 M - \frac{\lambda}{2} M^2 - \frac{\lambda'}{4} M^4 \tag{11.55}$$

and, consequently, Eq. (11.28) changes to

$$F = -(B_0 \mu) m + \left(\frac{3}{2} k_B T - \frac{1}{2} \lambda \mu^2 \right) m^2 - \left(\frac{1}{4} \lambda' \mu^4 - \frac{9}{20} k_B T \right) m^4$$
$$+ \left(\frac{297}{1050} k_B T \right) m^6. \tag{11.56}$$

A simple comparison of Eqs. (11.37) and (11.56) leads to

$$g_1 = B_0\mu \tag{11.57}$$
$$g_2 = \alpha(T - T_0),$$
$$g_4 = \alpha'(T_0' - T) \quad \text{for } T > T_0',$$
$$g_6 = \frac{297}{175} k_B T,$$

where

$$\alpha = 3k_B \quad \text{and} \quad T_0 = \frac{\lambda\mu^2}{3k_B}, \tag{11.58}$$

$$\alpha' = \frac{9}{5}k_B \quad \text{and} \quad T_0' = \frac{5}{9}\frac{\lambda'\mu^4}{k_B}. \tag{11.59}$$

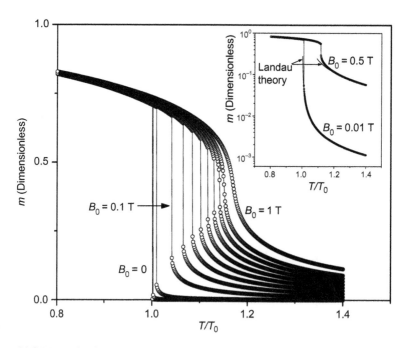

Figure 11.8 Magnetization (order parameter), as a function of temperature and under an external applied field B_0. Here, magnetization is given by the mean field Langevin equation (obtained via numerical calculation), considering the M^3 term to the mean field approximation. The inset emphasizes the agreement of Eq. (11.63) and the mean field Langevin equation. The full line is only a guide to the eyes.

From Eq. (11.42) it is easy to obtain the critical temperature T_c

$$T_c = \frac{T_0'}{(\Gamma_c - 1)} \left\{ \left(\frac{\Gamma_c T_0}{2 T_0'} - 1 \right) + \left[\left(\frac{\Gamma_c T_0}{2 T_0'} - 1 \right)^2 + (\Gamma_c - 1) \right]^{1/2} \right\}, \qquad (11.60)$$

where

$$\Gamma_c = \frac{176}{21} \approx 8.38. \qquad (11.61)$$

Analogously, from Eq. (11.44) (quite similar to Eq. (11.42)), it is possible to obtain T_{sh}, that, on its turn, is the same for T_c, i.e., Eq. (11.60); however, instead of Γ_c, we must consider

$$\Gamma_{sh} = \frac{44}{7} \approx 6.29. \qquad (11.62)$$

Note Figure 11.7 to see these temperatures and the thermal hysteresis.

Considering the influence of external force, we can obtain, from Eq. (11.53), the behavior of the order parameter, i.e., the magnetization, as a function of temperature and magnetic field. Thus,

$$m = \frac{\mu_B}{k_B} \frac{B_0 \mu}{3(T - T_0)}. \qquad (11.63)$$

See Figure 11.8 to note the influence of the external field on the supercooling temperature and the approximation above determined.

12 Molecular Magnetism

Molecular magnetism is a new research area and, in short, contains organic and metal-organic materials with low dimension and 3D character. Also considered as molecular magnets are the inorganic materials with low dimensional character (e.g., clusters and chains). Here, dimension is related to the magnetic lattice, while the crystal lattice for all of these materials is always 3D.

12.1 Zero-dimensional magnets

Zero-dimensional molecular magnets attract attention due to the huge number of potential applications, namely, high density data storage [37], quantum information and computation [38], magnetic refrigeration [39], spintronics [37], and photo-induced magnetism [40], that make these materials able to be used in devices.

However, a careful understanding of the fundamental questions behind these materials still needs attention, especially those related to some quantum behaviors (starting point for new technologies). In this sense, new magnetic phenomenon challenges the well-established concepts and then promises to give rise to revolutionary technologies. Thus, new materials still need to be synthesized and magnetically understood then to be used in those applications above mentioned (and other applications, of course).

12.1.1 Isotropic cases

12.1.1.1 Dimers

A dimeric unity (represented in Figure 12.1) can be written by isotropic Heisenberg Hamiltonian:

$$\mathcal{H}_{\text{hei}} = -J \vec{S}_1 \cdot \vec{S}_2, \tag{12.1}$$

where J is the exchange parameter (see Chapter 9 for further details). Eigenvalues and thermodynamic quantities are not simple to be obtained manually, and thus a computational routine can be useful. In this sense, the eigenvalues of energy and thermodynamic quantities here presented were obtained using the CARDAMOMO package [41–43], a computational routine to develop models on molecular magnetism (see Section 12.1.3 for further details).

Table 12.1 presents the eigenstates and eigenvalues E_i for this Hamiltonian. These energy levels are for $s_1 = s_2 = 5/2$. To obtain those for $s_1 = s_2 = 5/2 - 1/2 = 2$,

Fundamentals of Magnetism. http://dx.doi.org/10.1016/B978-0-12-405545-2.00012-6

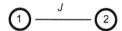

Figure 12.1 Isotropic dimer scheme.

Table 12.1 Eigenstates in the basis $|s, m_s\rangle$, degeneracy for zero magnetic field and energy eigenvalues for a dimeric unit $s_1 = s_2 = 5/2$. To obtain those for $s_1 = s_2 = 5/2 - 1/2 = 2$, remove the first line ($s = 5$). For $s_1 = s_2 = 2 - 1/2 = 3/2$, remove other line; down to $s_1 = s_2 = 1/2$, where there are only the two last lines ($s = 1$ and $s = 0$).

Label	Eigenstate $|s, m_s\rangle$	Degeneracy ($B_z = 0$)	Eigenvalue ($B_z = 0$)
E_5	$|5, m_s\rangle$	11	$-15J$
E_4	$|4, m_s\rangle$	9	$-10J$
E_3	$|3, m_s\rangle$	7	$-6J$
E_2	$|2, m_s\rangle$	5	$-3J$
E_1	$|1, m_s\rangle$	3	$-J$
E_0	$|0, m_s\rangle$	1	0

remove the first line ($s = 5$); and for $s_1 = s_2 = 2 - 1/2 = 3/2$, remove other line, down to $s_1 = s_2 = 1/2$, where there are only the two last lines ($s = 1$ and $s = 0$).

From the information given in Chapter 5, it is possible to obtain the thermodynamic quantities from the energy eigenvalues. Thus, the magnetic susceptibility (for $s = s_1 = s_2 = 5/2$) reads as

$$\chi = \mu_0 N_d \frac{2(g\mu_B)^2}{k_B T} \frac{\mathcal{N}}{Z}, \tag{12.2}$$

where

$$\mathcal{N} = e^{-E_1/t} + 5e^{-E_2/t} + 14e^{-E_3/t} + 30e^{-E_4/t} + 55e^{-E_5/t}, \tag{12.3}$$

$$Z = 1 + 3e^{-E_1/t} + 5e^{-E_2/t} + 7e^{-E_3/t} + 9e^{-E_4/t} + 11e^{-E_5/t}, \tag{12.4}$$

$t = k_B T$ and N_d is the number of dimers in the system. To obtain this quantity for $s_1 = s_2 = 5/2 - 1/2 = 2$, remove the last term of the numerator and denominator. For $s_1 = s_2 = 2 - 1/2 = 3/2$, remove other term of the numerator and denominator; down to $s_1 = s_2 = 1/2$, where

$$\frac{\mathcal{N}}{Z} = \frac{e^{-E_1/t}}{1 + 3e^{-E_1/t}}. \tag{12.5}$$

The same idea works to the specific heat without external magnetic field. For $s_1 = s_2 = 5/2$ it reads as

$$C_B = x^2 \frac{\mathcal{N}'}{Z^2} N_d k_B, \tag{12.6}$$

where

$$x = \frac{J}{k_B T},\tag{12.7}$$

$$\mathcal{N}' = (3e^x)\tag{12.8}$$
$$+ (45e^{3x} + 60e^{4x})$$
$$+ (252e^{6x} + 315e^{9x} + 525e^{7x})$$
$$+ (900e^{10x} + 1008e^{16x} + 2187e^{11x} + 2205e^{13x})$$
$$+ (2475e^{25x} + 2475e^{15x} + 6237e^{21x} + 6468e^{16x} + 7920e^{18x})$$

and Z is given by Eq. (12.4). For this case, to obtain the specific heat of other values of s, remove the last term of the denominator; however, for the numerator, remove each parenthesis (instead of each term). This operation returns, for $s_1 = s_2 = 1/2$:

$$\frac{\mathcal{N}'}{Z^2} = \frac{3e^x}{(1 + 3e^x)^2}.\tag{12.9}$$

Figure 12.2 summarizes these results and shows magnetization, magnetic suscepti-bility, and specific heat for an isotropic dimeric unity. Note these quantities are in real (laboratory) units.

Let us turn our attention to Table 12.1. To include the magnetic field B to the energy spectra, we only need to add the Zeeman term:

$$m_s g \mu_B B_z.\tag{12.10}$$

As discussed in Chapter 9, there is a correction to the Heisenberg Hamiltonian: the biquadratic term:

$$\mathcal{H}_{\text{biq}} = -j \left(\vec{S}_1 \cdot \vec{S}_2 \right)^2.\tag{12.11}$$

This energy is quite smaller than the Heisenberg term ($j/J \simeq 0.01$). Table 12.2 presents the energy levels of $1/2 \le (s_1 = s_2) \le 5/2$ dimers considering this cor-rection. Remember, these eigenvalues were obtained using the CARDAMOMO package (see Section 12.1.3).

12.1.1.2   Trimers

Consider the general trimer presented in Figure 12.3. From this figure, the Hamiltonian is:

$$\mathcal{H} = -J_1(\vec{S}_1 \cdot \vec{S}_3 + \vec{S}_2 \cdot \vec{S}_3) - J_2 \vec{S}_1 \cdot \vec{S}_2,\tag{12.12}$$

that can be written in terms of the sequential coupling of spins (see complements of this chapter). Thus,

$$\mathcal{H} = -\frac{J_1}{2} S^2 - \frac{(J_2 - J_1)}{2} S_{12}^2 + \text{const.},\tag{12.13}$$

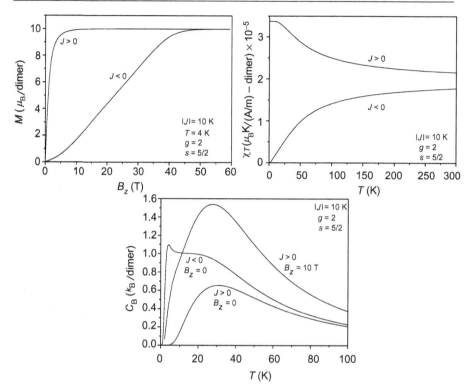

Figure 12.2 Magnetization M as a function of the magnetic field B, susceptibility χ times temperature T, i.e., χT, and specific heat under constant magnetic field C_B, as a function of temperature for a dimer ($s_1 = s_2$). Note these quantities are in real and measurable units. The quantities with applied magnetic field were obtained numerically.

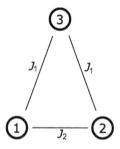

Figure 12.3 Scheme of a trimer. Note this trimer is isosceles and to make it linear, consider $J_2 = 0$. To be equilateral, consider $J_1 = J_2$.

Table 12.2 Energy spectra of dimers considering the bi-quadratic Hamiltonian correction to the Heisenberg one.

Label	Eigenstate $\lvert s, m_s \rangle$	Degeneracy ($B_z = 0$)	Eigenvalue ($B_z = 0$)
		$s_1 = s_2 = 5/2$	
E_5	$\lvert 5, m_s \rangle$	11	$-15J + 75j/2$
E_4	$\lvert 4, m_s \rangle$	9	$-10J + 75j$
E_3	$\lvert 3, m_s \rangle$	7	$-6J + 69j$
E_2	$\lvert 2, m_s \rangle$	5	$-3J + 87j/2$
E_1	$\lvert 1, m_s \rangle$	3	$-J + 33j/2$
E_0	$\lvert 0, m_s \rangle$	1	0
		$s_1 = s_2 = 2$	
E_4	$\lvert 4, m_s \rangle$	9	$-10J + 20j$
E_3	$\lvert 3, m_s \rangle$	7	$-6J + 36j$
E_2	$\lvert 2, m_s \rangle$	5	$-3J + 27j$
E_1	$\lvert 1, m_s \rangle$	3	$-J + 11j$
E_0	$\lvert 0, m_s \rangle$	1	0
		$s_1 = s_2 = 3/2$	
E_3	$\lvert 3, m_s \rangle$	7	$-6J + 9j$
E_2	$\lvert 2, m_s \rangle$	5	$-3J + 27/2\,j$
E_1	$\lvert 1, m_s \rangle$	3	$-J + 13/2\,j$
E_0	$\lvert 0, m_s \rangle$	1	0
		$s_1 = s_2 = 1$	
E_2	$\lvert 2, m_s \rangle$	5	$-3J + 3j$
E_1	$\lvert 1, m_s \rangle$	3	$-J + 3j$
E_0	$\lvert 0, m_s \rangle$	1	0
		$s_1 = s_2 = 1/2$	
E_1	$\lvert 1, m_s \rangle$	3	$-J + 1/2\,j$
E_0	$\lvert 0, m_s \rangle$	1	0

where

$$\vec{S}_{12} = \vec{S}_1 + \vec{S}_2 \quad \text{and} \quad \vec{S} = \vec{S}_{12} + \vec{S}_3. \tag{12.14}$$

Note the above Hamiltonian has linear terms on S_{12}^2 and S^2 and therefore is diagonal on the basis $\lvert s_{12}, s, m_s \rangle$. It has a significant importance, since we do not need to diagonalize the Hamiltonian by means of computational procedures; this basis change makes the Hamiltonian 12.12 diagonal. Note the trimer of Figure 12.3 is isosceles and considering $J_2 = 0$ it becomes linear; and equilateral considering $J_1 = J_2$. The above Hamiltonian has eigenstates and eigenvalues as those presented in Table 12.3.

From the information given in Chapter 5 and the energy spectra presented in Table 12.3, it is possible to obtain the thermodynamic quantities (see Section 12.1.3

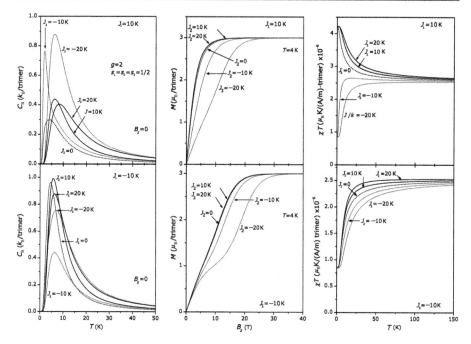

Figure 12.4 Specific heat at constant magnetic field C_B as a function of temperature (left), magnetization M as a function of magnetic field (center), and susceptibility times temperature χT as a function of temperature (right), for a trimer ($s_1 = s_2 = s_3 = 1/2$). The top figures have $J_1 > 0$, while those at the bottom have $J_1 < 0$. Dotted lines are configurations of J_1 and J_2 that have frustration, i.e., those with $J_2 < 0$. Dashed lines are for the linear system, i.e., $J_2 = 0$.

for details on how to obtain thermodynamic quantities automatically). The magnetic susceptibility for the trimer of Figure 12.3 and $s_1 = s_2 = s_3 = 1/2$ reads as

$$\chi = \mu_0 N_t \frac{(g\mu_B)^2}{2k_B T} \frac{\mathcal{N}}{Z}, \tag{12.15}$$

where N_t is the number of trimers in the system,

$$\mathcal{N} = 1 + e^{-E_1/t} + 10e^{-E_2/t}, \tag{12.16}$$

$$Z = 2 + 2e^{-E_1/t} + 4e^{-E_2/t}, \tag{12.17}$$

and $t = k_B T$. The energy spectra is given in Table 12.3.

The same idea works for the specific heat at zero magnetic field:

$$C_B = \frac{2}{(k_B T)^2} \frac{\mathcal{N}'}{Z^2} N_t k_B, \tag{12.18}$$

where

$$\mathcal{N}' = 4E_2^2 e^{-E_2/t} + 9J_1^2 e^{-(E_1+E_2)/t} + 2E_1^2 e^{-E_1/t} \tag{12.19}$$

and Z is given by Eq. (12.17). These results are presented in Figure 12.4.

Table 12.3 Eigenstate in the basis $|s_{12}, s, m_s\rangle$, degeneracy for zero magnetic field, and energy eigenvalues for a trimer unit $s_1 = s_2 = s_3 = 1/2$. See Figure 12.3.

Label	Eigenstates $\|s_{12}, s, m_s\rangle$	Degeneracy ($B_z = 0$)	Eigenvalues ($B_z = 0$)
E_2	$\|1, 3/2, m_s\rangle$	4	$-(J_1 + 2J_2)/2$
E_1	$\|1, 1/2, m_s\rangle$	2	$J_1 - J_2$
E_0	$\|0, 1/2, m_s\rangle$	2	0

Table 12.4 Configurations of J_1 and J_2 that lead to frustration.

	J_1	J_2	
$J_1 > 0$ triangle	+	+	*nonfrustrated*
	+	−	*frustrated*
$J_1 < 0$ triangle	−	+	*nonfrustrated*
	−	−	*frustrated*

Back to Table 12.3, to add the Zeeman contribution it is needed to include:

$$m_s g \mu_B B_z \tag{12.20}$$

to those zero-field eigenvalues.

An interesting discussion comes from this trimer model: geometric frustration, i.e., exchange interactions that lead to a conflict among the possible spin orientations. See Table 12.4. Note for either $J_1 > 0$ or $J_1 < 0$, only positive J_2 is possible; however, this last can assume negative values and therefore the system becomes frustrated. The dotted lines in Figure 12.4 represent frustrated configurations.

12.1.2 Anisotropic cases

This subsection deals with the magnetic susceptibility of anisotropic clusters, those with, for instance, local magnetocrystalline anisotropy or dipolar interactions (this last, left as an exercise to the reader). Here, all of the magnetic susceptibilities are normalized by the number of clusters and μ_0.

12.1.2.1 Local magnetocrystalline anisotropy

Let us start from the Hamiltonian that describes this anisotropy (see Eq. (9.69)):

$$\mathcal{H}_{ani} = \vec{S} \cdot \overleftrightarrow{\mathcal{D}}_{ani} \cdot \vec{S}, \tag{12.21}$$

where $\overleftrightarrow{\mathcal{D}}_{ani}$ is a tensor given by Eq. (9.70).

For the sake of simplicity, let us only consider the axial case, i.e., $D \neq 0$ and $E = 0$; and, as an example, $s = 3/2$. For other values of s, it is possible to follow the idea here presented, but, in any case, the eigenvalues of energy will be presented for $1 \leq s \leq 5/2$.

We can start from the Hamiltonian, in the matrix form, of Eq. (9.96) (with $E = 0$ and built with the help of the CARDAMOMO package—see Section 12.1.3):

$$
\mathcal{H} = D
\begin{pmatrix}
1 & 0 & 0 & 0 \\
0 & 1 & 0 & 0 \\
0 & 0 & -1 & 0 \\
0 & 0 & 0 & -1
\end{pmatrix}
\begin{array}{l}
\langle +3/2| \\
\langle -3/2| \\
\langle +1/2| \\
\langle -1/2|
\end{array}
\;,
\tag{12.22}
$$

where the spin operators were built up on the local basis $|s, m_s\rangle$ and the matrix ordered following the states $|m_s\rangle$ placed on the side of the above matrix. Now, to obtain the magnetic susceptibility, it is needed to add a Zeeman term (from Eq. (9.81)):

$$
\mu_B \vec{B} \cdot \overset{\leftrightarrow}{g} \cdot \vec{S} = \mu_B \sum_u \sum_v g_{uv} S_u B_v,
\tag{12.23}
$$

where $u, v = x, y, z$. Considering only the diagonal terms of the $\overset{\leftrightarrow}{g}$ tensor, the Zeeman contribution resumes as

$$
\mu_B \sum_u g_{uu} S_u B_u.
\tag{12.24}
$$

Note, for a polycrystal we need to consider (but it is an approximation), the magnetic field along u and then evaluate χ_u; then, the average magnetic susceptibility reads as

$$
\chi_m = \frac{1}{3}(\chi_x + \chi_y + \chi_z).
\tag{12.25}
$$

Let us start considering the magnetic field along x. Thus, the Zeeman term is (again, see Section 12.1.3 for further details on how to build this matrix automatically)

$$
\mathcal{H}_x = b_x
\begin{pmatrix}
0 & 0 & \sqrt{3}/2 & 0 \\
0 & 0 & 0 & \sqrt{3}/2 \\
\sqrt{3}/2 & 0 & 0 & 1 \\
0 & \sqrt{3}/2 & 1 & 0
\end{pmatrix},
\tag{12.26}
$$

where

$$
b_x = g_x B_x \mu_B.
\tag{12.27}
$$

Note the zero-field Hamiltonian is diagonal (Eq. (12.22)), while the Zeeman term is not. This situation allows the use of Perturbation Theory (see Complement 8.A), to obtain the eigenvalues of energy (see Table 12.5), and then the van Vleck susceptibility (see Section 8.3). Again, all of these steps can be evaluated with the CARDAMOMO package (see Section 12.1.3). Thus:

Table 12.5 Energy eigenvalues considering the local magnetocrystalline anisotropy for the axial case ($D \neq 0$ and $E = 0$) and the magnetic field along z and x directions. The eigenvalues for the case in which the magnetic field is along y are the same for the x case. Find in this table eigenvalues for $1 \leq s \leq 5/2$. $s = 1/2$ have no magnetocrystalline anisotropy, since the zero field Hamiltonian is null.

$|s, m_s\rangle$	$E_{m_s}^{(0)}$	$E_{m_s}^{(1)}(b_z)$	$E_{m_s}^{(2)}(b_z)$	$E_{m_s}^{(1)}(b_x)$	$E_{m_s}^{(2)}(b_x)$
			$s = 1$		
1	D	b_z	0	0	$b_x^2/2D$
-1	D	$-b_z$	0	0	$b_x^2/2D$
0	0	0	0	0	$-b_x^2/D$
			$s = 3/2$		
3/2	$2D$	$3b_z/2$	0	0	$3b_x^2/8D$
$-3/2$	$2D$	$-3b_z/2$	0	0	$3b_x^2/8D$
1/2	0	$b_z/2$	0	$-b_x$	$-3b_x^2/8D$
$-1/2$	0	$-b_z/2$	0	b_x	$-3b_x^2/8D$
			$s = 2$		
2	$4D$	$2b_z$	0	0	$b_x^2/3D$
-2	$4D$	$-2b_z$	0	0	$b_x^2/3D$
1	D	b_z	0	0	$7b_x^2/6D$
-1	D	$-b_z$	0	0	$7b_x^2/6D$
0	0	0	0	0	$-3b_x^2/D$
			$s = 5/2$		
5/2	$6D$	$5b_z/2$	0	0	$5b_x^2/16D$
$-5/2$	$6D$	$-5b_z/2$	0	0	$5b_x^2/16D$
3/2	$2D$	$3b_z/2$	0	0	$11b_x^2/16D$
$-3/2$	$2D$	$-3b_z/2$	0	0	$11b_x^2/16D$
1/2	0	$b_z/2$	0	$-3b_x/2$	$-b_x^2/D$
$-1/2$	0	$-b_z/2$	0	$3b_x/2$	$-b_x^2/D$

$$\chi_x = \mu_0 N \frac{(g_x \mu_B)^2}{4D} \frac{3 + 4x - 3e^{-2x}}{1 + e^{-2x}}, \qquad (12.28)$$

where

$$x = \frac{D}{k_B T} \qquad (12.29)$$

and N is the number of spins in the system.

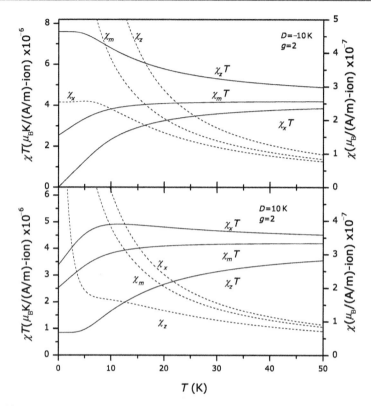

Figure 12.5 Susceptibility times temperature and susceptibility due to the local magnetocrystalline anisotropy. The subscript of each susceptibility represents the direction in which the magnetic field is along. χ_m represents the mean magnetic susceptibility.

Analogously, it is possible to execute the same procedure to the magnetic field along z. The eigenvalues are also given in Table 12.5 and then, the van Vleck susceptibility is

$$\chi_z = \mu_0 N \frac{(g_z \mu_B)^2}{4 k_B T} \frac{(1 + 9 e^{-2x})}{1 + e^{-2x}}. \tag{12.30}$$

It is easy to show that the energy eigenvalues (and, consequently, the magnetic susceptibility) are the same for the cases in which the magnetic field is along x and y. Finally, the average magnetic susceptibility is

$$\chi_m = \frac{1}{3}(\chi_z + 2\chi_x). \tag{12.31}$$

The magnetic susceptibility for the cases $D > 0$ and $D < 0$ is given in Figure 12.5. Note, considering a polycrystal, i.e., microcrystals along all of the directions, the average susceptibility, χ_m (Eq. (12.31)), is quite similar for both cases ($D > 0$ and $D < 0$). Thus, for polycrystals it is quite difficult to determine the sign of the D parameter.

On the other hand, note the huge difference for the case of single crystals; where, for instance, for $D > 0$, $\chi_z T$ decreases by decreasing temperature; while for $D < 0$, the contrary occurs.

To conclude this section, see Table 12.5, where the energy eigenvalues for the cases $s = 1$, $3/2$, 2, and $5/2$ are presented. Finally, find below the respective magnetic susceptibilities when the magnetic field is along x and z (remember $\chi_x = \chi_y$). For $s = 1$:

$$\chi_z = \mu_0 N 2 \frac{(g_z \mu_B)^2}{k_B T} \frac{e^{-x}}{1 + 2e^{-x}} \tag{12.32}$$

and

$$\chi_x = \mu_0 N 2 \frac{(g_x \mu_B)^2}{D} \frac{1 - e^{-x}}{1 + 2e^{-x}}. \tag{12.33}$$

For $s = 2$:

$$\chi_z = \mu_0 N 2 \frac{(g_z \mu_B)^2}{k_B T} \frac{e^{-x}(1 + 4e^{-3x})}{1 + 2e^{-x} + 2e^{-4x}} \tag{12.34}$$

and

$$\chi_x = \mu_0 N 2 \frac{(g_x \mu_B)^2}{3D} \frac{9 - 7e^{-x} - 2e^{-4x}}{1 + 2e^{-x} + 2e^{-4x}}. \tag{12.35}$$

Finally, for $s = 5/2$:

$$\chi_z = \mu_0 N \frac{(g_z \mu_B)^2}{4 k_B T} \frac{(1 + 9e^{-2x} + 25e^{-6x})}{1 + e^{-2x} + e^{-6x}} \tag{12.36}$$

and

$$\chi_x = \mu_0 N 2 \frac{(g_x \mu_B)^2}{3D} \frac{16 + 18x - 11e^{-2x} - 5e^{-6x}}{1 + e^{-2x} + e^{-6x}}. \tag{12.37}$$

It is quite important to stress that the zero-field Hamiltonian 12.21 for $s = 1/2$ is

$$\mathcal{H} = D \begin{pmatrix} 0 & 0 \\ 0 & 0 \end{pmatrix} \begin{matrix} \langle +1/2| \\ \langle -1/2| \end{matrix}, \tag{12.38}$$

and therefore $s = 1/2$ spin has no influence of local magnetocrystalline anisotropy.

12.1.2.2 Dipolar interaction

Evaluation of this problem is left as an exercise to the reader; considering a $s_1 = s_2 = 1/2$ dimer with an isotropic Heisenberg interaction added to a dipolar one.

12.1.3   Computational routine: the CARDAMOMO package

As mentioned before, the eigenvalues and thermodynamic quantities of the systems presented in this Chapter are easy to be obtained, however, hard to be evaluated manually. In this sense, a computational routine can be useful, and the CARDAMOMO package is one of these. We strongly recommend the specific literature [42–44] for a better understanding.

12.1.4   Modeling zero-dimensional molecular magnets

This subsection presents a qualitative description of how to develop models for zero-dimensional molecular magnets, i.e., clusters.

1. *Geometric configuration:* The first step is to obtain the crystallographic structure of the molecular magnet under consideration, with special attention to the magnetic centers (metallic and/or organic). Following, we must transfer the information of this structure (the whole crystal), to the geometrical configuration of spins. Based therefore on the bond angles and distances between magnetic ions, it is possible to associate one or more exchange interactions. Different angles/distances mean different exchange interactions (at least in first approximation).

2. *Hamiltonian:* From the geometric configuration of spins, it is possible then to write the Hamiltonian. A discussion on the physical meaning of some Hamiltonians is given in Chapter 9. Those contributions are spin Hamiltonians, i.e., depend only on the spins of the system.

3. *Eigenvalues:* There are several ways to do this task. It can be done from basis change or perturbation theory.

   3.a. Basis change: Spin Hamiltonian containing S_x and S_y are not diagonal on the local basis $|s_1, s_2, \ldots, s_i, m_{s_1}, m_{s_2}, \ldots, m_{s_i}\rangle$ (only z component), and then basis change can be a useful and powerful way to obtain eigenvalues. For some cases, there is a known coupled basis in which the Hamiltonian is diagonal and the Clebsch-Gordan coefficients for these two bases (local and coupled) can be used to change the basis of the Hamiltonian. There are two kinds of coupled basis: sequential and nonsequential. For the first case, the spins are summed up one-by-one sequentially. For instance, considering four spins (s_1, s_2, s_3, and s_4). It is possible to couple those as: $S_{12} = S_1 + S_2$, $S_{13} = S_{12} + S_3$, and $S_{14} = S = S_{13} + S_4$. Note that the spins are coupled sequentially. For the second case, the spins are coupled nonsequentially. Considering the spins above, it is possible to couple those as (other ways are possible): $S_{12} = S_1 + S_2$, $S_{34} = S_3 + S_4$, and $S = S_{12} + S_{34}$. See the Complements of this chapter for a further understanding on basis change.

   3.b. Perturbation Theory: This way to obtain the eigenvalues considers the zero-field Hamiltonian as the nonperturbed term (must be diagonal), and the Zeeman term as a perturbation (either diagonal or nondiagonal). Complement 8.A presents a brief survey on perturbation theory. From this approach, only the van Vleck susceptibility can be evaluated.

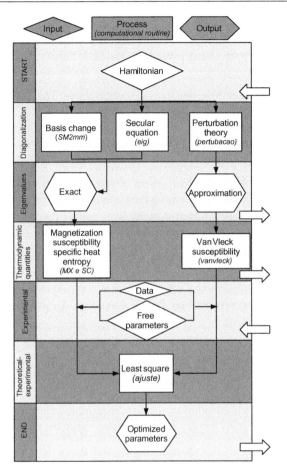

Figure 12.6 Fluxogram to develop models. Lozenges mean information needed to run the model, i.e., inputs. Rectangles mean internal evaluation, i.e., neither input nor output. Heptagons mean outputs. Some of these seven steps have in and out arrows (right-hand side of the diagram) and indicate that either input is required or output is available. Names between parentheses are the computational routines that execute the corresponding task into the CARDAMOMO package.

4. *Thermodynamic quantities:* The previous step gave us the energy spectra and at this point we are able to obtain the thermodynamic quantities, namely magnetization, magnetic susceptibility, specific heat, and entropy (see Chapter 5).

5. *Experimental data:* The analytical thermodynamic quantities obtained in the previous step can be compared with experimental data of interest. The free parameters to the fitting can be, for instance, the exchange parameter J of the isotropic Heisenberg Hamiltonian, the Lande factor g, and other quantities.

6. *Fitting to the experimental data:* At this level of the model, it is needed to use some least square procedure to obtain then the best fitting.

7. *Free parameters:* To conclude the model, the outputs are the optimized free parameters and a theoretical curve. With these values it is possible to understand the microscopic phenomena behind the magnetic behavior of a such molecular magnet.

Figure 12.6 summarizes the seven steps described above. The CARDAMOMO package (see Section 12.1.3) follows these steps.

12.1.5   Example and computational routine

As example of application of the procedure above described, let us consider the compound $Na_2Cu_5Si_4O_{14}$. It has a zig-zag chain of copper dimers and trimers, as shown in Figure 12.7. Further details concerning the material can be found in reference [45].

First, from the crystal structure it is possible to extract the magnetic structure, as depicted in Figure 12.8. The Hamiltonian then holds as

$$\mathcal{H} = -J_1(S_1 S_2 + S_2 S_3) - J_2 S_A S_B - J_3 S_4 S_5 - g\mu_B B S_z, \tag{12.39}$$

where $S_B = S_1 + S_2 + S_3$, $S_A = S_4 + S_5$, and $S = S_A + S_B$; this last corresponds to the total spin of the dimer-trimer pair. Let us also define $S_{B'} = S_1 + S_3$, and rewrite Eq. (12.39) in the following way:

$$\mathcal{H} = -\frac{J_1}{2}(S_B^2 - S_{B'}^2 - S_2^2) - \frac{J_2}{2}(S^2 - S_A^2 - S_B^2) - \frac{J_3}{2}(S_A^2 - S_4^2 - S_5^2) - g\mu_B B S. \tag{12.40}$$

Thus, we can see from the equation above that the proposed Hamiltonian can be diagonalized by means of basis change and, in this case, using the nonsequential coupled basis: $|s, s_A, s_B, s_{B'}\rangle$.

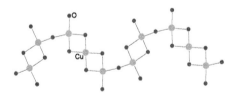

Figure 12.7 Crystal structure of the $Na_2Cu_5Si_4O_{14}$ compound, emphasizing the zig-zag copper chain. Those other ions in the chemical composition are omitted from the figure for the sake of clarity.

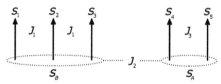

Figure 12.8 Schematic view of the dimer-trimer cluster, with the respective exchange interactions: J_1-intra-trimer, J_3-intra-dimer, and J_2-inter-trimer-dimer.

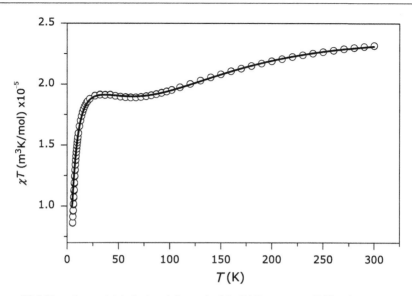

Figure 12.9 Experimental (circles) and theoretical (solid line) susceptibility times temperature χT for the compound $Na_2Cu_5Si_4O_{14}$. The solid line was obtained using the CARDAMOMO package. See text for further details.

The free parameters optimized (using the CARDAMOMO package) are: $J_1 = -254.8$ K, $J_2 = -9.6$ K, $J_3 = 31.7$ K, $g = 2.36$, and $\chi_{ti} = -1.9 \times 10^{-7}$ μ_B/FU–Oe. See Figure 12.9. Above, χ_{ti} is an added temperature independent contribution to the susceptibility due to either diamagnetism or Pauli paramagnetism.

In what concerns computational routine, it is recommended to see the CARDAMOMO package: a computational routine do develop models on molecular magnets. Further information can be found in the program website [42] and some articles [43,44].

12.2   One-dimensional magnets

One-dimensional molecular magnets have special importance: these are test-bench of theoretical models. These materials represent a genuine realization of magnetic chains that, in a first approximation, are magnetically disconnected. On the other hand, due to the evolution of the synthesis techniques, several new materials have been prepared and these require an understanding, based on those already established models.

It is important to stress that the magnetic susceptibility of chains, for most of the cases, has no analytical solution (some exceptions are the Ising model and classical Heisenberg chains), and the usual procedure is to obtain a numerical solution and then fit a polynomial expression to the numerical solution, to obtain the polynomial parameters. Then, it is possible to fit the polynomial solution to the experimental data, with some free parameters; usually the exchange interaction J of the Heisenberg Hamiltonian, the Lande factor g, and other parameters.

Figure 12.10 Uni-metallic regular chain scheme.

There are several possible combinations for a magnetic chain. It is possible to find both, either regular or irregular space between metallic centers (and, consequently, a sort of strength of exchange interaction between the first neighbors); chains with only one (uni-metallic) or two (duo-metallic), types of metallic center (of course, more metallic centers are also possible to be found); and much more rich systems.

12.2.1 Uni-metallic regular chain

First, let us define a uni-metallic regular chain: obviously, only one type of metallic ion and a regular space between those through the chain. Figure 12.10 clarifies this idea. The Hamiltonian of this case is:

$$\mathcal{H} = -J \sum_{i=1}^{N} \vec{S}_i \cdot \vec{S}_{i+1} + g\mu_B \sum_{i=1}^{N} \vec{S}_i \cdot \vec{B}. \tag{12.41}$$

The magnetic susceptibility from this Hamiltonian has no analytical solution to the limit $N \to \infty$; however, numerical solution is possible by considering the boundary condition

$$\vec{S}_{N+1} = \vec{S}_1, \tag{12.42}$$

i.e., a ring.

For $s = 1/2$, the numerical solution was found by Bonner and Fisher [45], in 1964, for $J < 0$, and reads as

$$\chi = N\mu_0 \frac{(g\mu_B)^2}{k_B T} \frac{\mathcal{N}}{\mathcal{D}}, \tag{12.43}$$

where

$$\mathcal{N} = 0.25 + 0.074975x + 0.075235x^2, \tag{12.44}$$
$$\mathcal{D} = 1 + 0.9931x + 0.172135x^2 + 0.757825x^3, \tag{12.45}$$

and

$$x = \frac{|J|}{k_B T}. \tag{12.46}$$

Later in 1982, Meyer and coworkers [47] found a similar result, however, for $s = 1$. For this case, $\mathcal{N}$ and $\mathcal{D}$ above assume:

$$\mathcal{N} = 2 + 0.0194x + 0.777x^2 \tag{12.47}$$

and

$$\mathcal{D} = 3 + 4.346x + 3.232x^2 + 5.834x^3. \tag{12.48}$$

Further on the results, in 1964, Fisher [48] obtained an analytical solution for the magnetic susceptibility considering classical spins for the Hamiltonian in Equation 12.41. The susceptibility is therefore:

$$\chi = N\mu_0 \frac{(g\mu_B)^2}{3k_B T} s(s+1) \frac{1+u}{1-u}, \tag{12.49}$$

where

$$u = \coth(y) - \frac{1}{y} \tag{12.50}$$

and

$$y = \frac{Js(s+1)}{k_B T}. \tag{12.51}$$

This case works better for $s \geq 5/2$.

Finally, a particular case is the Ising model; also described by the Hamiltonian in Equation 12.41, however, considering only the z projection of the $1/2$ spin. When the applied magnetic field is along z direction, the magnetic susceptibility reads as:

$$\chi_{zz} = N\mu_0 \frac{(g_z\mu_B)^2}{4k_B T} e^{J/2k_B T}, \tag{12.52}$$

an analytical result obtained by Kramers and Wannier in 1941 [49]. In addition, when the magnetic field is applied along the x direction (i.e., perpendicular to the spin direction), the magnetic susceptibility reads as:

$$\chi_{xx} = N\mu_0 \frac{(g_x\mu_B)^2}{2J} \left[\tanh\left(\frac{J}{2k_B T}\right) + \left(\frac{J}{4k_B T}\right) \text{sech}^2\left(\frac{J}{4k_B T}\right) \right]. \tag{12.53}$$

The above result was obtained by Fisher [50].

The behavior of the magnetic susceptibilities above presented is depicted in Figure 12.11.

12.2.2 Duo-metallic regular chain

This case is quite analogous to before, however, instead of a unique kind of ion, the system has two different types of ions: A and B, equally spaced through the chain. Figure 12.12 clarifies this scenario.

The Hamiltonian of this case is

$$\mathcal{H} = -J \sum_{i=1}^{N} \vec{S}_i \cdot \vec{S}_{i+1} + \mu_B \sum_{i=1}^{N} g_i \vec{S}_i \cdot \vec{B}, \tag{12.54}$$

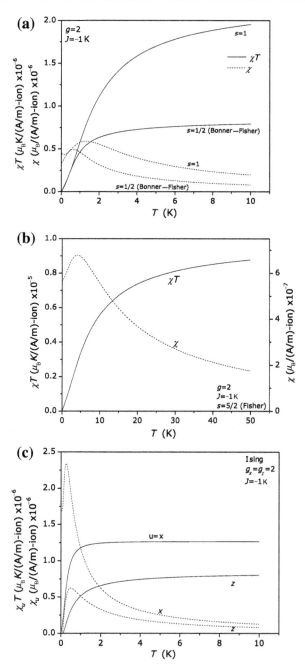

Figure 12.11 Magnetic susceptibility χ and χT as a function of temperature, considering $g = 2$ and $J = -1$ K, for a ring chain and several different cases. (a) $s = 1/2$ (Bonner–Fisher) and $s = 1$. (b) Classical limit (works for $s \geq 5/2$). The present case is $s = 5/2$. (c) Ising model: there is only the z component of the $1/2$ spin. u means the direction of the applied magnetic field.

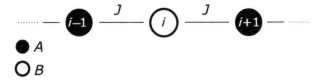

Figure 12.12 Duo-metallic regular chain scheme.

where

$$S_{\text{odd}} = S_A \qquad g_{\text{odd}} = g_A$$
$$S_{\text{even}} = S_B \qquad g_{\text{even}} = g_B$$

As before, there is no analytical solution to this system; only numerical one, considering a ring boundary condition:

$$\vec{S}_{N+1} = \vec{S}_1. \tag{12.55}$$

Analogously to before, Drillon and coworkers [50], in 1989, obtained a numerical solution to the magnetic susceptibility of a duo-metallic chain with $s_A = 1/2$ and $s_B = 1$, with antiferromagnetic interaction between first neighbors ($J < 0$):

$$\chi = N_{AB}\mu_0 \frac{(g\mu_B)^2}{k_B T} \frac{\mathcal{N}}{\mathcal{D}}, \tag{12.56}$$

where

$$\mathcal{N} = 11 - 7.231x + 2.81693x^2 - 0.0341468x^3, \tag{12.57}$$

$$\mathcal{D} = 12 + 0.697190x + 1.29663x^2, \tag{12.58}$$

and

$$x = \frac{|J|}{k_B T}. \tag{12.59}$$

Above, N_{AB} is the number of $A - B$ pairs. They considered $g_A = g_B = g$. Figure 12.13 presents the behavior of the magnetic susceptibility above presented. Note there is a minimum in χT and at low temperatures this quantity diverges. This minimum, at $T_{\min}$, for several values of s_B (up to $5/2$), is presented in Table 12.6. Note if we know this minimum from the experimental data, it is possible to obtain the exchange parameter J.

Following these ideas, it is important to cite the work of Georges and coworkers [52], that proposed a friendly expression to the magnetic susceptibility, for $s_A = 1/2$ and $1 \le s_B \le 5/2$. This analytical expression was fitted to the numerical results obtained by them [53]. Georges proposal is therefore:

$$\chi = N_{AB}\mu_0 \frac{2(g\mu_B)^2}{3k_B T} f(x), \tag{12.60}$$

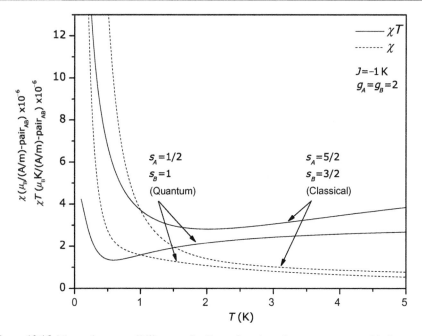

Figure 12.13 Magnetic susceptibility χ and χT as a function of temperature, considering $g_A = g_B = 2$ and $J = -1$ K, for a duo-metallic chain and two different cases: (i) $s_A = 1/2$ and $s_B = 1$ (two quantum spins); and (ii) $s_A = 5/2$ and $s_B = 3/2$ (two almost-classical spins).

Table 12.6 Values of temperature $T_{\min}$ where the minimum in χT occurs, for a regular duo-metallic chain with spins $s_A = 1/2$ and $1 \leq s_B \leq 5/2$. These results consider $g_A = g_B$.

$(s_A - s_B)$	$k_B T_{\min}/J$
$(1/2 - 1)$	0.58
$(1/2 - 3/2)$	1.19
$(1/2 - 2)$	2.10
$(1/2 - 5/2)$	2.90

where

$$f(x) = ax^\beta + C' \exp(-bx) \tag{12.61}$$

and

$$x = \frac{|J|}{k_B T}. \tag{12.62}$$

Above, a, b, and β are constants fitted to the numerical results [52] and

$$C' = C'_A + C'_B = \frac{1}{2}\left[s_A(s_A + 1) + s_B(s_B + 1)\right] \tag{12.63}$$

Table 12.7 Parameter values of Eq. (12.61), fitted to the numerical results of reference [52].

s_B	a	β	C'	b
1	0.1146	1.69	1.375	0.811
3/2	0.5756	1.80	2.250	0.882
2	1.8120	1.78	3.375	0.948
5/2	4.1249	1.72	4.750	1.013

is proportional to the Curie constant. Again, this model considers $g_A = g_B$. Those values of a, b, β, and C' are given in Table 12.7.

Stepping forward, let us now review other model: the classical one. Drillon and coworkers [54], in 1983, obtained an analytical expression for a duo-metallic chain of classical spins s_A and s_B. They found:

$$\chi = N_{AB}\mu_0 \frac{\mu_B^2}{3k_B T}\left(\bar{g}^2\frac{1+u}{1-u} + \bar{\bar{g}}^2\frac{1-u}{1+u}\right), \tag{12.64}$$

where

$$u = \coth(x) - \frac{1}{x} \tag{12.65}$$

and

$$x = \frac{\bar{J}}{k_B T}. \tag{12.66}$$

In addition:

$$\bar{g} = \frac{1}{2}(\bar{g}_A + \bar{g}_B), \quad \bar{\bar{g}} = \frac{1}{2}(\bar{g}_A - \bar{g}_B) \tag{12.67}$$

and

$$\bar{g}_A = g_A\sqrt{s_A(s_A+1)}, \quad \bar{g}_B = g_B\sqrt{s_B(s_B+1)}. \tag{12.68}$$

Finally,

$$\bar{J} = J\sqrt{s_A(s_A+1)s_B(s_B+1)}. \tag{12.69}$$

Figure 12.13 presents the behavior of the magnetic susceptibility above presented.

12.2.3   Uni-metallic irregular chain

This case deals with a chain with only one kind of magnetic ions, not equally spaced through the chain; i.e., there are two exchange parameters J and αJ. It is interesting, because α can assume either positive or negative values. For positive values, the chain has an antiferro-antiferro-antiferro (AF-AF-AF) periodicity; while for negative values of α, the chain has an AF-F-AF periodicity (F means ferromagnetic). Figure 12.14 helps us to see this system.

Figure 12.14 Uni-metallic irregular chain scheme.

The Hamiltonian of this case is

$$\mathcal{H} = -J \sum_{i=1}^{N} (\vec{S}_{i-1} \cdot \vec{S}_i + \alpha \vec{S}_i \cdot \vec{S}_{i+1}) + g\mu_B \sum_{i=1}^{N} \vec{S}_i \cdot \vec{B}. \tag{12.70}$$

First, let us focus our attention to a $s = 1/2$ chain with AF-AF-AF periodicity. This case was first developed, in 1968, by Duffy and Barr [55], and then, in 1981, Hall and coworkers [56] fitted an analytical expression to the numerical results of Duffy and Barr. Then, Hall et al. obtained a polynomial in α and

$$x = \frac{|J|}{k_B T}. \tag{12.71}$$

Hall susceptibility is then given by

$$\chi = N\mu_0 \frac{(g\mu_B)^2}{k_B T} \frac{\mathcal{N}}{\mathcal{D}}. \tag{12.72}$$

It works only for $J < 0$ and $0 \leq \alpha \leq 1$, i.e., an AF-AF-AF chain. Table 12.8 has the values of $\mathcal{N}$ and $\mathcal{D}$.

Table 12.8 Values of $\mathcal{N}$ and $\mathcal{D}$ of Eq. (12.72), for different ranges of α. Here, $x = |J|/k_B T$.

	$0.4 < \alpha \leq 1$
$\mathcal{N}$	$(0.25) +$ $(-0.068475 + 0.13194\alpha)x +$ $(0.0042563 - 0.031670\alpha + 0.12278\alpha^2 - 0.29943\alpha^3 + 0.21814\alpha^4)x^2$
$\mathcal{D}$	$(1) +$ $(0.035255 + 0.65210\alpha)x +$ $(-0.00089418 - 0.10209\alpha + 0.87155\alpha^2 - 0.18472\alpha^3)x^2 +$ $(0.045230 - 0.0081910\alpha + 0.83234\alpha^2 - 2.6181\alpha^3 + 1.92813\alpha^4)x^3$
	$0 \leq \alpha \leq 0.4$
$\mathcal{N}$	$(0.25) +$ $(-0.062935 + 0.11376\alpha)x +$ $(0.0047778 - 0.033268\alpha + 0.12742\alpha^2 - 0.32918\alpha^3 + 0.25203\alpha^4)x^2$
$\mathcal{D}$	$(1) +$ $(0.053860 + 0.70960\alpha)x +$ $(-0.00071302 - 0.10587\alpha + 0.54883\alpha^2 - 0.20603\alpha^3)x^2 +$ $(0.047193 - 0.0083778\alpha + 0.87256\alpha^2 - 2.7098\alpha^3 + 1.9798\alpha^4)x^3$

Table 12.9 Values of $\mathcal{N}$ and $\mathcal{D}$ of Eq. (12.74), for different values of α. Here, $y = k_B T/|J|$.

	$-2 \le \alpha \le 0$
$\mathcal{N}$	$y^3 + 5y^2 - y + 0.05$
$\mathcal{D}$	$y^4 +$ $(5.2623 - 0.33021|\alpha|)y^3 +$ $(0.44977 - 0.99235|\alpha| - 0.00882|\alpha|^2 + 0.15482|\alpha|^3)y^2 +$ $(0.18948 + 0.36766|\alpha| + 0.51001|\alpha|^2 - 0.27958|\alpha|^3)y +$ $(0.28438 - 0.16750|\alpha| - 0.18725|\alpha|^2 + 0.09375|\alpha|^3)$
	$-8 \le \alpha \le -2$
$\mathcal{N}$	$y^3 +$ $5y^2 +$ $(18.49536 - 6.13263|\alpha| + 1.63541|\alpha|^2 - 0.11494|\alpha|^3)y +$ $(-1.47602 + 0.23810|\alpha| - 0.03943|\alpha|^2 + 0.00185|\alpha|^3)$
$\mathcal{D}$	$y^4 +$ $(5.31957 - 0.25252|\alpha|)y^3 +$ $(20.1290 - 7.98424|\alpha| + 1.82750|\alpha|^2 - 0.11683|\alpha|^3)y^2 +$ $(-2.69685 + 2.71648|\alpha| - 0.31049|\alpha|^2 + 0.00834|\alpha|^3)y +$ $(5.11208 - 2.47824|\alpha| + 0.45708|\alpha|^2 - 0.02687|\alpha|^3)$

Further results were obtained by Borras and coworkers [56], in 1994, where these authors considered an AF-F-AF chain, i.e., $J < 0$ and $\alpha < 0$. They also fitted the numerical result; however with a polynomial in α and

$$y = \frac{k_B T}{|J|}. \tag{12.73}$$

Borras susceptibility is

$$\chi = N\mu_0 \frac{(g\mu_B)^2}{4|J|} \frac{\mathcal{N}}{\mathcal{D}}, \tag{12.74}$$

where $\mathcal{N}$ and $\mathcal{D}$ are given in Table 12.9.

Those susceptibilities, obtained by Hall and Borras, are valid only above a certain temperature, given below:

- $T > 0.25|J|$ for $0 \le \alpha \le 1$
- $T > 0.09|J|$ for $-2 \le \alpha \le 0$
- $T > 0.15|J|$ for $-5 \le \alpha \le -2$
- $T > 0.20|J|$ for $-8 \le \alpha \le -5$

Finally, Figure 12.15 summarizes the behavior of those susceptibilities, for several values of α. We also present, for the sake of completeness, isolated dimers (corresponds to $\alpha = 0$) and the Bonner–Fisher case (corresponds to $\alpha = 1$).

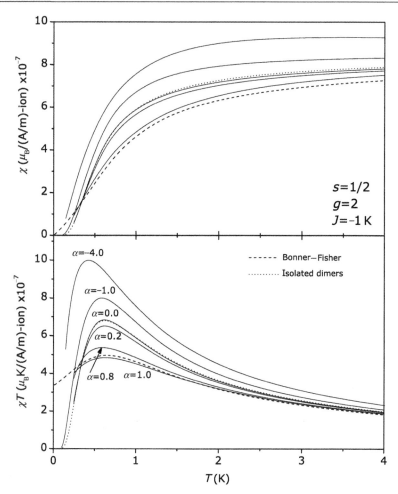

Figure 12.15 Susceptibility (top) and susceptibility times temperature (bottom), as a function of temperature for an irregular chain and the cases: $\alpha = -4.0, -1.0, 0.0, 0.2, 0.8, 1.0$. These results are only valid for $J < 0$. We also presented: (i) Bonner–Fisher (dashed line: $\alpha = 1$ limit) and (ii) isolated dimers (dotted line: $\alpha = 0$ limit).

12.2.4   Example and computational routine

We consider as an example a Mn^{2+} regular chain found in $[Mn(H_2O)_2(HBTC) \cdot (H_2O)]$, where BTC = 1, 2, 4-benzenetricarboxylate. Further details on the material can be found in the literature [58]. These ions are Mn^{2+} and therefore $s = 5/2$.

Fitting of the model to the experimental data is given in Figure 12.16. The optimized parameters are $J = -2.5$ K and $g = 1.9$.

In what concerns computational routine, it is recommended to see the CARDAMOMO package: a computational routine to develop models on molecular magnets. Further information can be found in the program website [48] and some articles [43,44].

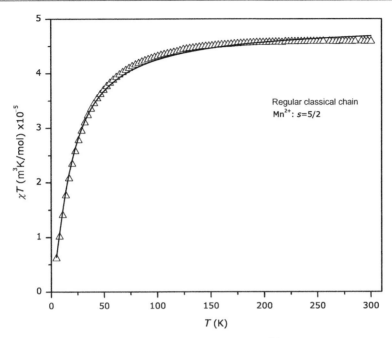

Figure 12.16 Example of a classical regular chain of $Mn^{2+}(s = 5/2)$, fitted with the CARDAMOMO package [42–44].

12.3   Single-molecule magnets

Some 0D magnets, i.e., clusters, due to the huge intra-cluster exchange interaction and local magnetocrystalline anisotropy, have a magnetic ground state, i.e., $m_s \neq 0$. As a consequence, this class of materials has some properties of conventional magnets and then justifies the name in which are known: single-molecule magnets or, in short, SMM.

For most of the cases, these clusters have some ions (around 10), and then the dimension of the Hilbert space is quite large and therefore difficult to be solved analytically. To overcome this problem, it is common to consider only the ground multiplet s (that works only at low values of temperature), and, consequently, the following Hamiltonian:

$$\mathcal{H} = \vec{S} \cdot \overset{\leftrightarrow}{\mathcal{D}}_{ani} \cdot \vec{S} + \mu_B \vec{B} \cdot \overset{\leftrightarrow}{g} \cdot \vec{S}. \tag{12.75}$$

To simplify, it is possible to assume (i) only the axial contribution of $\overset{\leftrightarrow}{\mathcal{D}}_{ani}$, i.e., $E = 0$, (ii) the magnetic field along z, (iii) the isotropy of the $\overset{\leftrightarrow}{g}$ tensor and, finally (iv) the term S^2 as a constant. Thus, the eigenvalues of energy are (see Eq. (9.96)):

$$E(m_s) = Dm_s^2 + g\mu_B B m_s, \tag{12.76}$$

where $g = g_{zz}$ and $B = B_z$.

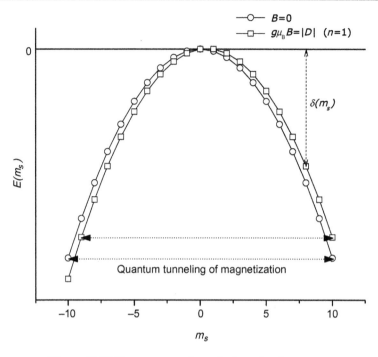

Figure 12.17 Energy spectra for a single-molecule magnet.

First, let us consider the zero field eigenvalues. It is easy to see that if $D > 0$ the ground state is $m_s = 0$ and then there is no SMM. On the other hand, if $D < 0$ the ground states are $m_s = \pm s$ and for this case, it is a SMM. In accordance with Eq. (9.84), to satisfy the $D < 0$ condition, the following inequality must be true: $2D_{zz} < D_{xx} + D_{yy}$. These D_{aa} components, in accordance with Eq. (9.70), depend on the projection of the angular momentum along a direction, that, on its turn, depends on the crystal symmetry. Thus, negative D strongly depends on the crystal symmetry. Finally, it is possible to rewrite the energy spectra of a SMM as:

$$E(m_s) = -|D|m_s^2 + g\mu_B B m_s. \tag{12.77}$$

To summarize this scenario, Figure 12.17 shows the energy spectra as a function of m_s. Note there is an energy barrier between the two possible ground states. The height of this barrier is:

$$\delta(m_s) = E(0) - E(m_s) = |D|m_s^2 - g\mu_B B m_s. \tag{12.78}$$

To flip the spin, i.e., change from the $m_s = -s$ state to $m_s = s$ (or the contrary), there are three possibilities. The first one is to apply a high enough magnetic field. Considering that for this case the condition $E(s) > E(s - 1)$ must be satisfied, it is

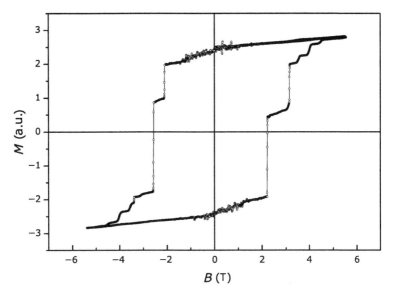

Figure 12.18 Magnetization curve of Mn12-ac. These steps correspond to the quantum tunneling of the magnetization. Measurement done by Prof. Angelo Gomes (UFRJ-Brazil).

easy to see that the needed magnetic field is:

$$B > \frac{|D|}{g\mu_B}(2s - 1) \tag{12.79}$$

and, as will be discussed below, for a real SMM (Mn12-ac), this magnetic field is of the order of 5 T. The second way to flip the spin is a thermal energy (at zero magnetic field), and therefore:

$$T > \frac{|D|s^2}{k_B}. \tag{12.80}$$

Analogously to the first case, for a standard SMM (Mn12-ac), this temperature is of the order of 40 K. Finally, the third possibility is the quantum tunneling of the magnetization. When a m_s state tunes other state $-m_s + n$, there is a probability to tunneling this energy barrier. The resonance occurs for:

$$E(m_s) = E(-m_s + n), \tag{12.81}$$

where n is the nth level above (or below) the level $-m_s$. From the condition above, the resonance condition is:

$$B = n\frac{|D|}{g\mu_B}. \tag{12.82}$$

For instance, for a transition from $m_s = s$ to $m_s = -s + 1$, a magnetic field of the order of 0.5 T, considering a standard SMM (Mn12-ac), is needed.

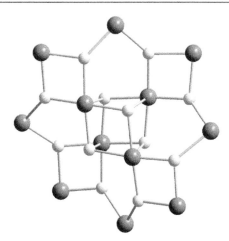

Figure 12.19 Mn12-ac cluster. Mn ions are the bigger circles and O ions are the smaller circles. Note four Mn ions at the center of the structure and eight around. Figure provided by Antonio dos Santos (ORNL-USA).

A famous SMM is the Mn12-ac. This material, see Figure 12.19, has four Mn^{4+} ($s_1 = 3/2$) at the center of the cluster and eight Mn^{3+} ($s_2 = 2$) around. Note the Hilbert space is quite large: $(2s_1 + 1)^4(2s_2 + 1)^8 = 10^8$, and therefore it is quite difficult to obtain analytical solution for all of the eigenvalues. However, experimental results of magnetization at high magnetic field and AC susceptibility verify that the ground state is a multiplet $s = 10$ [59,60]; and then allow the use of the Hamiltonian for only this ground state (at low temperature regime, i.e., below the energy scale of the intra-cluster exchange interactions). Figure 12.18 presents the magnetization curve at low temperatures, in which the quantum tunneling of magnetization can be seen; and from this result the D parameter can be obtained: $D \approx 0.5$ K [60] (see Eq. (12.82)).

Complements

12.A   Ising model: analytic solution

In 1925, Ising proposed a model: a spins' chain with first neighbor exchange interaction J, where these spins can only assume two values ± 1. The Hamiltonian is therefore:

$$\mathcal{H} = -J \sum_{i=1}^{N} \sigma_i \sigma_{i+1} - b \sum_{i=1}^{N} \sigma_i, \qquad (A.1)$$

where

$$b = g_z \mu_B B_z. \qquad (A.2)$$

Further below, a connection of the above Hamiltonian and that one in Eq. (12.41) will be provided; it means, rewrite the above Hamiltonian into the spin space. Following, the partition function reads as:

$$Z = \sum_{\{\sigma_i\}} \exp \left\{ \beta \sum_{i=1}^{N} \left[J \sigma_i \sigma_{i+1} + \frac{b}{2} \left(\sigma_i + \sigma_{i+1} \right) \right] \right\}, \tag{A.3}$$

where $\sigma_i \rightarrow (\sigma_i + \sigma_{i+1})/2$. There are some ways to go further [60] and here we will use the matrix form, proposed by Kramers and Wannier [49]. Thus, the partition function is:

$$Z = \sum_{\{\sigma_i\}} \prod_{i=1}^{N} \langle \sigma_i | \mathcal{P} | \sigma_{i+1} \rangle, \tag{A.4}$$

where

$$\langle \sigma_i | \mathcal{P} | \sigma_{i+1} \rangle = \exp \left\{ \beta \left[J \sigma_i \sigma_{i+1} + \frac{b}{2} \left(\sigma_i + \sigma_{i+1} \right) \right] \right\} \tag{A.5}$$

is known as *transfer matrix*. The index into brackets {} means that the sum runs over all of the spins σ_i, i.e., a sum (that can assume only two values ± 1), for each spin. Thus, we can expand it as:

$$Z = \sum_{\sigma_1} \sum_{\sigma_2} \cdots \sum_{\sigma_N} \langle \sigma_1 | \mathcal{P} | \sigma_2 \rangle \langle \sigma_2 | \mathcal{P} | \sigma_3 \rangle \cdots \langle \sigma_N | \mathcal{P} | \sigma_1 \rangle. \tag{A.6}$$

Note the boundary condition

$$\sigma_{N+1} = \sigma_1 \tag{A.7}$$

has been considered, i.e., a ring. Considering now the closure relation of the basis of each spin σ_i:

$$\sum_{\sigma_i} |\sigma_i\rangle \langle \sigma_i| = \mathbb{I}, \tag{A.8}$$

where $\mathbb{I}$ is the identity matrix, then the partition function reads as:

$$Z = \sum_{\sigma_1} \langle \sigma_1 | \mathcal{P}^N | \sigma_1 \rangle = \mathrm{Tr} \left\{ \mathcal{P}^N \right\}. \tag{A.9}$$

Note the ring condition (Eq. (A.7)) is necessary for this simplification and the partition function is the trace of the transfer matrix. Let us now see this matrix in more detail.

The matrix elements in lines (*bras*) are not dependent on those in rows (*kets*) and therefore it is possible to assume always two independent values (± 1), and, for this case, it is easy to built the matrix:

$$\mathcal{P} = \begin{array}{cc} & \begin{array}{cc} |+\rangle & |-\rangle \end{array} \\ \begin{pmatrix} e^{\beta(J+b)} & e^{-\beta J} \\ e^{-\beta J} & e^{\beta(J-b)} \end{pmatrix} & \begin{array}{c} \langle +| \\ \langle -| \end{array} \end{array}. \tag{A.10}$$

Note $\mathcal{P}$ is not diagonal, but it is possible to make it diagonal without losing information. Thus, to diagonalize the transfer matrix, let us use the secular equation:

$$\det \left(\mathcal{P} - \lambda \mathbb{I} \right) = 0, \tag{A.11}$$

where the two possible eigenvalues are:

$$\lambda_\pm = e^{\beta J}\left[\cosh(\beta b) \pm \sqrt{\sinh^2(\beta b) + e^{-4\beta J}}\right]. \tag{A.12}$$

Now:

$$P = \begin{pmatrix} \lambda_+ & 0 \\ 0 & \lambda_- \end{pmatrix} \Rightarrow P^N = \begin{pmatrix} \lambda_+^N & 0 \\ 0 & \lambda_-^N \end{pmatrix} \tag{A.13}$$

and then:

$$Z = \mathrm{Tr}\left\{P^N\right\} = \lambda_+^N + \lambda_-^N. \tag{A.14}$$

After this evaluation, here is the partition function in a closed and simple form. With the partition function, it is possible to obtain all of the thermodynamic quantities. Let us start with the Helmholtz free energy:

$$F = -k_B T \ln(Z), \tag{A.15}$$

that then assumes

$$\begin{aligned} F &= -\frac{1}{\beta} \ln\left[\lambda_+^N + \lambda_-^N\right] \\ &= -\frac{1}{\beta} \ln\left\{\lambda_+^N\left[1 + \left(\frac{\lambda_-}{\lambda_+}\right)^N\right]\right\} \\ &= -\frac{1}{\beta} \ln\left(\lambda_+^N\right) - \frac{1}{\beta} \ln\left[1 + \left(\frac{\lambda_-}{\lambda_+}\right)^N\right]. \end{aligned} \tag{A.16}$$

However, $\lambda_+ > \lambda_-$ for the thermodynamic limit, i.e., for $N \to \infty$, and then:

$$\left.\left(\frac{\lambda_-}{\lambda_+}\right)^N\right|_{N\to\infty} \to 0. \tag{A.17}$$

The free energy assumes thus:

$$F = -\frac{N}{\beta} \ln(\lambda_+). \tag{A.18}$$

The magnetization is given by:

$$M_u = -\frac{\partial F}{\partial B_u}, \tag{A.19}$$

but remember $b = g_z \mu_B B_z$, and then:

$$M_z = N g_z \mu_B \frac{\sinh(\beta b)}{\sqrt{\sinh^2(\beta b) + e^{-4\beta J}}}. \tag{A.20}$$

Considering now $\sigma = \sigma_z$, i.e., σ of the Ising Hamiltonian is the z component of the Pauli matrix, it is possible to rewrite these results into the spin space, by means of $J \to J/4$ and $B_z \to B_z/4$. Thus, the magnetization reads as:

$$M_z = N g_z \mu_B \frac{\sinh (\beta b/4)}{\sqrt{\sinh^2 (\beta b/4) + e^{-\beta J}}}. \qquad (A.21)$$

Note that

$$M(B_z = 0, T) = 0, \qquad (A.22)$$

i.e., for any value of $T > 0$, there is no spontaneous magnetization and the system is paramagnetic.

Finally, to obtain the magnetic susceptibility, we need the limit of low magnetic field. Thus:

$$M_z \xrightarrow{B_z \to 0} \frac{1}{4} \beta (g_z \mu_B)^2 B_z e^{\beta J/2} \qquad (A.23)$$

and then the magnetic susceptibility reads as:

$$\chi_{zz} = N \mu_0 \frac{(g_z \mu_B)^2}{4 k_B T} e^{J/2 k_B T}. \qquad (A.24)$$

This result was first obtained by Kramers and Wannier [48].

12.B   Addition of N spins

For the sake of simplicity, let us consider only four spins; but, of course, the idea here presented works for N spins. Thus, consider four spins: $s_1 = s_2 = s_3 = s_4 = 1/2$; these can be added either sequentially or nonsequentially. Starting from the sequential case, consider:

$$\begin{aligned} \vec{S}_{12} &= \vec{S}_1 + \vec{S}_2, \\ \vec{S}_{13} &= \vec{S}_{12} + \vec{S}_3, \\ \vec{S} = \vec{S}_{14} &= \vec{S}_{13} + \vec{S}_4. \end{aligned} \qquad (B.25)$$

From Eq. (3.33), the constraints are:

$$\begin{aligned} |s_1 - s_2| &\leq s_{12} \leq s_1 + s_2, \\ |s_{12} - s_3| &\leq s_{13} \leq s_{12} + s_3, \\ |s_{13} - s_4| &\leq s \leq s_{13} + s_4 \end{aligned} \qquad (B.26)$$

and then it is possible to build the coupled basis. First, s_{12} assumes two possible values: 1 and 0. Following, to obtain s_{13}, it is necessary to add s_3 to each possible value of s_{12}. For $s_{12} = 1$, two possible values of s_{13} arise: 3/2 and 1/2; and, from $s_{12} = 0$, only one value: 1/2. To obtain s_{14}, the same idea must be followed. Figure 12.20a

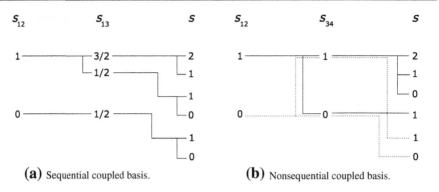

(a) Sequential coupled basis.   **(b)** Nonsequential coupled basis.

Figure 12.20 Addition of four 1/2 spins to build the coupled basis (sequential and nonsequential).

summarizes this procedure. Now, to obtain the basis, just follow the continuous lines from the right to the left, and each element of the last column starts a new state in the basis $|s_{12}, s_{13}, s, m_s\rangle$, where m_s is the projection of the s multiplet along z. Thus, the states are:

$$|1, 3/2, 2, m_s\rangle,$$
$$|1, 3/2, 1, m_s\rangle,$$
$$|1, 1/2, 1, m_s\rangle,$$
$$|1, 1/2, 0, m_s\rangle,$$
$$|0, 1/2, 1, m_s\rangle,$$
$$|0, 1/2, 0, m_s\rangle. \tag{B.27}$$

From now one, consider the basis coupled nonsequentially. One example is:

$$\vec{S}_{12} = \vec{S}_1 + \vec{S}_2, \tag{B.28}$$
$$\vec{S}_{34} = \vec{S}_3 + \vec{S}_4,$$
$$\vec{S} = \vec{S}_{12} + \vec{S}_{34}.$$

These new vectors also must follow the constraint of Eq. (3.33). To build the states, the procedure above described must be used. Figure 12.20b presents the diagram for this sum; below the states, written in the basis $|s_{12}, s_{34}, s, m_s\rangle$:

$$|1, 1, 2, m_s\rangle \tag{B.29}$$
$$|1, 1, 1, m_s\rangle$$
$$|1, 1, 0, m_s\rangle$$
$$|1, 0, 1, m_s\rangle$$
$$|0, 1, 1, m_s\rangle$$
$$|0, 0, 0, m_s\rangle$$

Finally, the CARDAMOMO package handles easily the construction of these bases (see Section 12.1.3).

Part Four

Appendices

Appendix A
Useful Mathematical Functions

This appendix describes some mathematical functions and their properties; those used in this book.

First, the PolyLogarithm function is defined as

$$Li_n(z) = \frac{1}{\Gamma(n)} \int_0^\infty \frac{x^{n-1}}{z^{-1}e^x - 1} dx, \quad \forall Re(n) > 0 \tag{A.1}$$

and its derivative as

$$\frac{\partial}{\partial z} Li_n(z) = \frac{1}{z} Li_{n-1}(z). \tag{A.2}$$

For small values of z, more precisely $|z| < 1$, this function reads as

$$Li_n(z) \approx \sum_{k=1}^\infty \frac{z^k}{k^n} = z + 2^{-n}z^2 + 3^{-n}z^3 + \cdots \tag{A.3}$$

In addition, for $z \to \infty$:

$$Li_n(z) \approx -\frac{[\ln(-z)]^n}{n!} + \frac{(-1)^{n-1}}{z} + 2 \sum_{k=1}^{n/2} \frac{Li_{2k}(-1)[\ln(-z)]^{n-2k}}{(n-2k)!}. \tag{A.4}$$

Some specific values are useful for this book:

$$Li_n(0) = 0, \tag{A.5}$$
$$Li_n(1) = \zeta(n), \tag{A.6}$$
$$Li_n(-1) = \zeta(n)(2^{1-n} - 1), \tag{A.7}$$

where ζ is the Riemann Zeta function, defined as

$$\zeta(n) = \frac{1}{\Gamma(n)} \int_0^\infty \frac{x^{n-1}}{e^x - 1} dx, \quad \forall Re(n) > 1. \tag{A.8}$$

Again, specific values are also useful for this book, specially

$$\zeta(2) = \frac{\pi^2}{6}. \tag{A.9}$$

Finally, the Gamma function is also used in this book, and reads as

$$\Gamma(n) = \int_0^\infty x^{n-1} e^{-x}\, dx, \quad \forall Re(n) > 0, \tag{A.10}$$

where specific values are

$$\Gamma(1/2) = \sqrt{\pi}, \quad \Gamma(3/2) = \frac{1}{2}\sqrt{\pi}, \quad \Gamma(5/2) = \frac{3}{4}\sqrt{\pi}, \quad \Gamma(7/2) = \frac{15}{8}\sqrt{\pi}. \tag{A.11}$$

Also used in this book is the factorial function written in terms of Gamma function, as below

$$n! = \Gamma(n+1). \tag{A.12}$$

Appendix B
Exercises

Chapter 1

1. Verify that the magnetic field H, far from the source, due to a ring (Eq. (1.6)), is the same as the dipolar field due to a permanent magnet (Eq. (1.8)).
2. Consider an uniformly magnetized sphere of radius R and magnetization M along the z axis. Obtain the demagnetization factor for this case, using for this purpose Eq. (1.18).

Chapter 2

1. Considering equivalent gauges $\{\vec{A}, \phi\}$ and $\{\vec{A}', \phi'\}$, given by Eqs. (2.A.13) and (2.A.14), shows that these potentials create the same magnetic field and electric field $\{\vec{B}, \vec{E}\}$, i.e.,

$$\vec{\nabla} \times \vec{A}' = \vec{\nabla} \times \vec{A} = \vec{B}, \tag{B.1}$$

$$-\vec{\nabla}\phi' - \frac{\partial \vec{A}'}{\partial t} = -\vec{\nabla}\phi - \frac{\partial \vec{A}}{\partial t} = \vec{E}. \tag{B.2}$$

Chapter 3

1. Write the operators J_x, J_y, and J_z on the matrix form, considering $j = 1$.
2. Obtain the Zeeman energy due to a single electron, considering only the spin contribution.

Chapter 4

1. Prove the following:

$$C_B = C_M + T \left[\frac{\partial M}{\partial T} \Big|_B \right]^2 \left[\frac{\partial M}{\partial B} \Big|_T \right]^{-1}, \tag{B.3}$$

where C_B and C_M are the specific heat with constant magnetic field and magnetization, respectively.

2. Prove the following:

$$\frac{C_B}{C_M} = \frac{\frac{\partial M}{\partial B}\big|_T}{\frac{\partial M}{\partial B}\big|_S}, \tag{B.4}$$

where C_B and C_M are the specific heat with constant magnetic field and magnetization, respectively.

Chapter 5

1. Considering the mean number of particles as a sum of the mean occupancy
number of each level, i.e.,

$$N = \sum_k \bar{n}_k \tag{B.5}$$

obtain the mean occupancy number $\bar{n}_k$ for those three different statistics (FD,
BE, and MB).

Chapter 6

1. Evaluate the density of states for a 1D and 2D Fermions gas.
2. Evaluate the *Grand* canonic potential Φ of an electron gas in the high temperature
limit (i.e., Maxwell–Boltzmann framework), using, however, Eq. (5.74) instead
of Eq. (5.72) into Eq. (6.21). Recover Eq. (6.41).
3. Show that for high temperature the fugacity tends to zero, i.e., $T \to \infty \Rightarrow z = e^{\mu\beta} \to 0$.

Chapter 7

1. Obtain the amplitude of the oscillations found on the magnetization of a diamagnetic gas under high magnetic field, i.e., the de Haas–van Alphen effect.
2. Consider the hydrogen atom discussed in Complement 3.A and evaluate the
diamagnetic susceptibility and magnetization of a 1s electron.
3. Derive the magnetization and magnetic susceptibility for the low magnetic field
and high temperature regime (Eqs. (7.47) and (7.48)), but starting from Eq. (7.57).
4. Evaluate the first term of the integral of Eq. (7.28) to obtain the result of Eq. (7.29).

Chapter 8

1. Is $C_M = 0$ expected? Verify this result considering the thermodynamic relation-ship between C_B and C_M

$$C_M = C_B - T \left[\frac{\partial M}{\partial T} \Big|_B \right]^2 \left[\frac{\partial M}{\partial B} \Big|_T \right]^{-1}. \tag{B.6}$$

2. Verify, using the relation

$$C_M = T \frac{\partial S}{\partial T} \Big|_M, \tag{B.7}$$

   that the specific heat at constant magnetization C_M is zero. Since it is not possible to do it analytically for the whole range of $x = b/t$, do it for the $x \to 0$ limit for both, quantum and classical cases.
3. Show that Brillouin function $B_j(x)$ recovers the Langevin function $L(x)$ for the $j \to \infty$ limit.
4. Prove that the total magnetization of an electron gas with spins is given by $M = (N_\Uparrow - N_\Downarrow)\mu_B$.
5. Evaluate the Pauli susceptibility, i.e., the paramagnetic contribution to the sus-ceptibility at low temperature limits ($\epsilon_F \gg k_B T$). However, instead of using Eq. (8.112), start from $q_\Uparrow(T, B, z)$ and $q_\Downarrow(T, B, z)$ (Eqs. (8.105) and (8.109), respectively), evaluate the corresponding *Grand* canonical potential $\Phi(T, B, z) = -\frac{1}{\beta}q(T, B, z)$ and, finally, the total magnetization $M = M_\Uparrow + M_\Downarrow$.
6. Obtain the non-perturbed energy, as well as the first- and second-order energy corrections of the Hamiltonians presented below.

   (a) *Nondegenerated case:*

$$\mathcal{H}_0 = \begin{pmatrix} 1 & 0 & 0 \\ 0 & 3 & 0 \\ 0 & 0 & -2 \end{pmatrix}, \quad \mathcal{W} = \begin{pmatrix} 0 & 1 & 0 \\ 1 & 0 & 0 \\ 0 & 0 & 1 \end{pmatrix}. \tag{B.8}$$

   (b) *Degenerated case:*

$$\mathcal{H}_0 = \begin{pmatrix} 1 & 0 & 0 \\ 0 & 1 & 0 \\ 0 & 0 & -2 \end{pmatrix}, \quad \mathcal{W} = \begin{pmatrix} 0 & 1 & 0 \\ 1 & 0 & 1 \\ 0 & 1 & 1 \end{pmatrix}. \tag{B.9}$$

7. Obtain χ_0 (Eq. (8.B.18)) and show that this term ($j = 0$) is responsible for the first term (24) of Eq. (8.B.22).

Chapter 9

1. Prove that symmetric (antisymmetric) spatial wave functions lead to antiparallel (parallel) alignment of the spins.
2. Obtain the Hamiltonian of a $s = 1/2$ spin with local magnetocrystalline anisotropy.

Chapter 11

1. Show that the maximum at ξ_{+-} goes to zero at T_0.

Chapter 12

1. Consider a dimer with $s_1 = s_2 = 1/2$ and write the following basis for this system: $|m_{s1}, m_{s2}\rangle$ and $|s, m_s\rangle$, where $|s_1 - s_2| \leqslant s \leqslant s_1 + s_2$.
2. Derive the magnetic susceptibility due to a $s_1 = s_2 = 1/2$ dimer with an isotropic Heisenberg interaction added to a dipolar one.

Appendix C
Solution of Exercises

Chapter 1

1. The magnetic field due to a permanent magnet is given by the dipolar field (Eq. (1.8)). Considering that the magnetic moment points along the z axis and our interesting point is also on this axis (then $\vec{r} = z\hat{k}$), it is possible to obtain

$$H = \frac{\mu}{2\pi z^3}.$$ (C.1)

Considering the magnetic field due to the ring (Eq. (1.6)), far from the ring ($a \ll z$), we must write

$$
\begin{aligned}
H &= \frac{I}{2} \frac{a^2}{z^3} \left(1 + \frac{a^2}{z^2}\right)^{-3/2} \\
&\approx \frac{I}{2} \frac{a^2}{z^3}.
\end{aligned}
$$ (C.2)

We know that the magnetic moment is $\mu = I\mathcal{A}$ (where $\mathcal{A}$ is the area of the ring), and then we get the same result of the permanent magnet:

$$H = \frac{\mu}{2\pi z^3}.$$ (C.3)

2. The magnetization is then $\vec{M} = M_0\hat{k}$ and a unitary vector normal to the sphere surface is

$$\hat{n} = \cos\phi \sin\theta\,\hat{i} + \sin\phi \sin\theta\,\hat{j} + \cos\theta\,\hat{k},$$ (C.4)

where $0 \le \theta \le \pi$ and $0 \le \phi \le 2\pi$ are, respectively, the zenith and azimuth angles of the spherical coordinates.

On this coordinates, the surface element can be written as

$$dS = r^2 \sin\theta\, d\theta\, d\phi$$ (C.5)

and then, from Eq. (1.18), the magnetic field inside the sphere is

$$
\vec{H}_{\mathrm{d}} = \frac{1}{4\pi} \int_0^\pi \int_0^{2\pi} (M_0 \cos\theta) \frac{R\hat{r}}{R^3} R^2 \sin\theta \, d\phi \, d\theta
$$

$$
= \frac{1}{4\pi} \int_0^\pi \int_0^{2\pi} (M_0 \cos\theta) \left[\cos\phi \sin\theta \hat{i} + \sin\phi \sin\theta \hat{j} \right.
$$

$$
\left. + \cos\theta \hat{k} \right] \sin\theta \, d\phi \, d\theta, \tag{C.6}
$$

where, for the sphere, $\hat{r} = \hat{n}$. Since $\int_0^{2\pi} \cos\phi \, d\phi = \int_0^{2\pi} \sin\phi \, d\phi = 0$, only the component along the z axis survives and then

$$
\vec{H}_{\mathrm{d}} = \frac{2\pi}{4\pi} M_0 \underbrace{\int_0^\pi \cos^2\theta \sin\theta \, d\theta}_{2/3} \hat{k}
$$

$$
= \frac{1}{3} M_0 \hat{k}. \tag{C.7}
$$

Thus, the demagnetization factor for the studied system is $N_{\mathrm{d}} = 1/3$.

Chapter 2

1. Equivalent gauges $\{\vec{A}, \phi\}$ and $\{\vec{A}', \phi'\}$ are given by (from Eqs. (2.A.13) and (2.A.14))

$$
\vec{A}'(\vec{r}, t) = \vec{A}(\vec{r}, t) + \vec{\nabla} f(\vec{r}, t), \tag{C.8}
$$

$$
\phi'(\vec{r}, t) = \phi(\vec{r}, t) - \frac{\partial}{\partial t} f(\vec{r}, t), \tag{C.9}
$$

where $f(\vec{r}, t)$ is any scalar field. Thus, for the magnetic field it reads as

$$
\vec{\nabla} \times \vec{A}' = \vec{\nabla} \times (\vec{A} + \vec{\nabla} f) \tag{C.10}
$$

$$
= \vec{\nabla} \times \vec{A} + \vec{\nabla} \times (\vec{\nabla} f)
$$

$$
= \vec{\nabla} \times \vec{A}
$$

$$
= \vec{B},
$$

since the rotational of the gradient of any scalar field is null, i.e., $\vec{\nabla} \times (\vec{\nabla} f) = 0$. For the electric field it is

$$
-\vec{\nabla}\phi' - \frac{\partial \vec{A}'}{\partial t} = -\vec{\nabla}\left(\phi - \frac{\partial f}{\partial t}\right) - \frac{\partial}{\partial t}\left(\vec{A} + \vec{\nabla} f\right)
$$

$$
= -\vec{\nabla}\phi - \frac{\partial \vec{A}}{\partial t}
$$

$$
= \vec{E}. \tag{C.11}
$$

We proved therefore that equivalent gauges $\{\vec{A}, \phi\}$ and $\{\vec{A}', \phi'\}$ given by Eqs. (2.A.13) and (2.A.14) can produce the same magnetic field and electric field $\{\vec{B}, \vec{E}\}$.

Chapter 3

1. Let us obtain the operators J_x, J_y, and J_z on the matrix form, for $j = 1$. For this case, there are three possibilities for m_j: -1, 0, and 1. Starting from J_z, this one is diagonal on the basis $|j, m_j\rangle$ (see Eq. (3.8)), where the matrix elements are given by

$$\langle j', m'_j | J_z | j, m_j \rangle = m_j \hbar \delta_{j',j} \delta_{m'_j, m_j}. \tag{C.12}$$

In what concerns J_x and J_y, we must obtain these from J_+ and J_-, that are not diagonal on the basis $|j, m_j\rangle$. The matrix elements of J_+ and J_- can then be written as (see Eq. (3.11))

$$\langle j', m'_j | J_\pm | j, m_j \rangle = \sqrt{(j \mp m_j)(j \pm m_j + 1)} \hbar \delta_{j',j} \delta_{m'_j, m_j \pm 1}. \tag{C.13}$$

From the above elements, it is possible to write J_z, J_+, and J_- on the matrix form

$$
J_z = \hbar
\begin{array}{ccc}
|1, +1\rangle & |1, 0\rangle & |1, -1\rangle \\
\end{array}
\left(
\begin{array}{ccc}
+1 & 0 & 0 \\
0 & 0 & 0 \\
0 & 0 & -1 \\
\end{array}
\right)
\begin{array}{c}
\langle 1, +1| \\
\langle 1, 0| \\
\langle 1, -1| \\
\end{array}
\tag{C.14}
$$

$$
J_+ = \hbar\sqrt{2}
\begin{pmatrix}
0 & 1 & 0 \\
0 & 0 & 1 \\
0 & 0 & 0 \\
\end{pmatrix}
\quad \text{and} \quad
J_- = \hbar\sqrt{2}
\begin{pmatrix}
0 & 0 & 0 \\
1 & 0 & 0 \\
0 & 1 & 0 \\
\end{pmatrix}.
\tag{C.15}
$$

Finally, considering Eq. (3.12) and the above matrices, we can write J_x and J_y

$$
J_x = \frac{\hbar}{\sqrt{2}}
\begin{pmatrix}
0 & 1 & 0 \\
1 & 0 & 1 \\
0 & 1 & 0 \\
\end{pmatrix},
\quad
J_y = \frac{\hbar i}{\sqrt{2}}
\begin{pmatrix}
0 & -1 & 0 \\
1 & 0 & -1 \\
0 & 1 & 0 \\
\end{pmatrix}.
\tag{C.16}
$$

2. From a similar fashion as before, it is possible to obtain the matrix form of the S_z operator for a single electron (i.e., $s = 1/2$). It reads as

$$
S_z = \frac{\hbar}{2}
\begin{pmatrix}
1 & 0 \\
0 & -1 \\
\end{pmatrix}
\tag{C.17}
$$

written on the basis $|s, m_s\rangle$. Note m_s assumes two possible values: $+1/2$ and $-1/2$. Let us simplify the notation and write the two possible states as: $| \uparrow \rangle = |1/2, +1/2\rangle$ and $| \downarrow \rangle = |1/2, -1/2\rangle$. The spin eigenvalues are

$$S_z| \uparrow \rangle = \frac{\hbar}{2}| \uparrow \rangle, \tag{C.18}$$

$$S_z| \downarrow \rangle = -\frac{\hbar}{2}| \downarrow \rangle. \tag{C.19}$$

The Zeeman energy for a spin only contribution is

$$\mathcal{H}_{S,L} = \frac{\mu_B}{\hbar} 2\vec{S} \cdot \vec{B} \tag{C.20}$$

and, considering $\vec{B} = B\hat{k}$, the above Hamiltonian resumes as

$$\mathcal{H}_{S,L} = \frac{\mu_B}{\hbar} 2S_z B. \tag{C.21}$$

Thus, since the Hamiltonian is linear with respect to the S_z operator, it is also diagonal on the working basis. The possible energies are then

$$E_\uparrow = +\mu_B B = E_\Downarrow, \tag{C.22}$$

$$E_\downarrow = -\mu_B B = E_\Uparrow. \tag{C.23}$$

Note the single leg arrows ($\uparrow$ and $\downarrow$) mean the direction of the spin vector and not the associated magnetic moment (represented as double leg arrows: $\Uparrow$ and $\Downarrow$), that points in opposition (see Eq. (3.37)).

Chapter 4

1. First, we need to consider an explicit dependence of the entropy on the magnetization M and temperature T:

$$S = S(M, T) \tag{C.24}$$

and then obtain the differential form:

$$T\, dS = T \left. \frac{\partial S}{\partial M} \right|_T dM + T \left. \frac{\partial S}{\partial T} \right|_M dT, \tag{C.25}$$

where the second term of the above equation is $C_M\, dT$ (see Eq. (4.46)). Analogously, we can also define the entropy as a function of the magnetic induction B and temperature T:

$$S = S(B, T) \tag{C.26}$$

and then:

$$T\, dS = T \left. \frac{\partial S}{\partial B} \right|_T dB + T \left. \frac{\partial S}{\partial T} \right|_B dT, \tag{C.27}$$

where the second term of the above equation is $C_B \, dT$ (see Eq. (4.47)).
After a simple comparison of Eqs. (C.25) and (C.27), we get

$$T \left. \frac{\partial S}{\partial M} \right|_T dM + C_M \, dT = T \left. \frac{\partial S}{\partial B} \right|_T dB + C_B \, dT. \tag{C.28}$$

To proceed, we need further considerations. It is reasonable to assume

$$M = M(B, T) \tag{C.29}$$

and then

$$dM = \left. \frac{\partial M}{\partial B} \right|_T dB + \left. \frac{\partial M}{\partial T} \right|_B dT. \tag{C.30}$$

Thus, inserting Eq. (C.30) into Eq. (C.28) leads to

$$T \left. \frac{\partial S}{\partial M} \right|_T \left\{ \left. \frac{\partial M}{\partial B} \right|_T dB + \left. \frac{\partial M}{\partial T} \right|_B dT \right\} + C_M \, dT = T \left. \frac{\partial S}{\partial B} \right|_T dB + C_B \, dT. \tag{C.31}$$

The terms on dT are

$$T \left. \frac{\partial S}{\partial M} \right|_T \left. \frac{\partial M}{\partial T} \right|_B + C_M = C_B \tag{C.32}$$

and the terms on dB are

$$T \left. \frac{\partial S}{\partial M} \right|_T \left. \frac{\partial M}{\partial B} \right|_T = T \left. \frac{\partial S}{\partial B} \right|_T. \tag{C.33}$$

Considering one Maxwell Relation (from the thermodynamic square):

$$\left. \frac{\partial S}{\partial B} \right|_T = \left. \frac{\partial M}{\partial T} \right|_B \tag{C.34}$$

and Eq. (C.33), it is possible to show that

$$\left. \frac{\partial S}{\partial M} \right|_T = \frac{\left. \frac{\partial M}{\partial T} \right|_B}{\left. \frac{\partial M}{\partial B} \right|_T}. \tag{C.35}$$

Finally, insert Eq. (C.35) into Eq. (C.32) is the last step to obtain our aim:

$$C_B = C_M + T \left[\left. \frac{\partial M}{\partial T} \right|_B \right]^2 \left[\left. \frac{\partial M}{\partial B} \right|_T \right]^{-1}. \tag{C.36}$$

2. In the last exercise we considered the entropy as a function of the magnetization and temperature

$$S = S(M, T). \tag{C.37}$$

Now, we still consider this assumption and let us take the inverse function:

$$M = M(S, T). \tag{C.38}$$

Thus, the differential form is

$$dM = \left.\frac{\partial M}{\partial S}\right|_T dS + \left.\frac{\partial M}{\partial T}\right|_S dT \tag{C.39}$$

and considering $dM = 0$ we get

$$\left.\frac{\partial S}{\partial T}\right|_M = -\frac{\left.\frac{\partial M}{\partial T}\right|_S}{\left.\frac{\partial M}{\partial S}\right|_T}. \tag{C.40}$$

Analogously to the last exercise, let us consider the entropy as below

$$S = S(B, T) \tag{C.41}$$

and its inverse:

$$B = B(S, T) \tag{C.42}$$

and then, the differential form:

$$dB = \left.\frac{\partial B}{\partial S}\right|_T dS + \left.\frac{\partial B}{\partial T}\right|_S dT. \tag{C.43}$$

Again, considering $dB = 0$ we obtain

$$\left.\frac{\partial S}{\partial T}\right|_B = -\frac{\left.\frac{\partial B}{\partial T}\right|_S}{\left.\frac{\partial B}{\partial S}\right|_T}. \tag{C.44}$$

From the definition of specific heat (Eqs. (4.46) and (4.47)), we know that

$$\frac{C_B}{C_M} = \frac{T \left.\frac{\partial S}{\partial T}\right|_B}{T \left.\frac{\partial S}{\partial T}\right|_M} \tag{C.45}$$

and then inserting Eqs. (C.40) and (C.44) into the above equation leads to

$$\frac{C_B}{C_M} = \frac{T \left.\frac{\partial S}{\partial T}\right|_B}{T \left.\frac{\partial S}{\partial T}\right|_M} = \frac{\left.\frac{\partial B}{\partial T}\right|_S \left.\frac{\partial M}{\partial S}\right|_T}{\left.\frac{\partial B}{\partial S}\right|_T \left.\frac{\partial M}{\partial T}\right|_S} \tag{C.46}$$

and then, after a simple evaluation, our final result becomes

$$\frac{C_B}{C_M} = \frac{\left.\frac{\partial M}{\partial B}\right|_T}{\left.\frac{\partial M}{\partial B}\right|_S}. \tag{C.47}$$

Chapter 5

1. From Eq. (5.65), we know that

$$N = -\frac{\partial \Phi}{\partial \mu} = k_B T \frac{\partial q}{\partial \mu} \tag{C.48}$$

considering, as defined:

$$\Phi = -\frac{1}{\beta} q. \tag{C.49}$$

Thus, we need $q = \ln \mathcal{Z}$ for each statistics. From Eq. (5.72) we have

$$q^{FD} = \sum_{k=1}^{+\infty} \ln \left(1 + z \, e^{-\beta \epsilon_k}\right) \tag{C.50}$$

and then

$$N^{FD} = \sum_{k=1}^{+\infty} \frac{1}{z^{-1} e^{\beta \epsilon_k} + 1}, \tag{C.51}$$

where we can obtain

$$\bar{n}_k^{FD} = \frac{1}{z^{-1} e^{\beta \epsilon_k} + 1}. \tag{C.52}$$

Analogously to before, from Eq. (5.73) we know that

$$q^{BE} = -\sum_{k=1}^{+\infty} \ln \left(1 - z \, e^{-\beta \epsilon_k}\right) \tag{C.53}$$

and then

$$\bar{n}_k^{BE} = \frac{1}{z^{-1} e^{\beta \epsilon_k} - 1}. \tag{C.54}$$

Finally, from Eq. (5.74)

$$q^{MB} = \sum_{k=1}^{+\infty} z \, e^{-\beta \epsilon_k} \tag{C.55}$$

and then

$$\bar{n}_k^{MB} = \frac{1}{z^{-1} e^{\beta \epsilon_k}}. \tag{C.56}$$

Figures C.1 and C.2 present these equations.

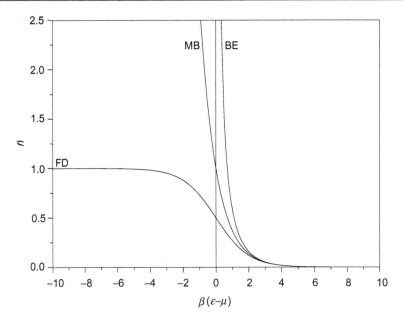

Figure C.1 Mean occupancy number for FD, BE, and MB statistics.

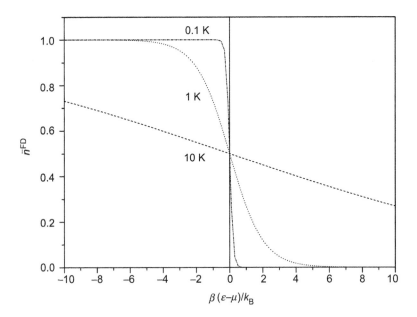

Figure C.2 Mean occupancy number for FD statistics, for different values of temperature.

Chapter 6

1. From Eq. (6.14) we know that

$$g(E) = \bar{g}\frac{dN_k}{dE} = \bar{g}\frac{dN_k}{dk}\frac{dk}{dE}, \tag{C.57}$$

where $\bar{g} = 2s+1$ was included "by hand" and takes into account the multiplicity due to the particle spin (for electron, $\bar{g} = 2$). Then, from Eq. (6.5)

$$\frac{dk}{dE} = \left(\frac{2m}{\hbar^2}\right)^{1/2}\frac{1}{2}E^{-1/2}. \tag{C.58}$$

Now, we need to find

$$N_k = \frac{V_k}{V_e}, \tag{C.59}$$

where V_k and V_e are volumes in the reciprocal space (the one contained by the vector $\vec{k}$ and the elemental one, respectively). Thus, for the 2D case,

$$V_e = \left(\frac{2\pi}{L}\right)^2 \quad \text{and} \quad V_k = \pi k^2 \tag{C.60}$$

and then

$$g(E) = \bar{g}\left(\frac{2m}{\hbar^2}\right)\frac{L^2}{4\pi}. \tag{C.61}$$

Analogously, for the 1D case,

$$V_e = \frac{2\pi}{L} \quad \text{and} \quad V_k = 2k \tag{C.62}$$

and then

$$g(E) = \frac{\bar{g}}{2}\left(\frac{2m}{\hbar^2}\right)^{1/2}\frac{L}{\pi}E^{-1/2}. \tag{C.63}$$

Note the dependence of these density of states on the energy. For the 3D case, $g(E)$ depends on $E^{1/2}$; for the 2D case, there is no dependence of the density of states on the energy; and, finally, for 1D case, it depends on $E^{-1/2}$.

2. Considering the Maxwell–Boltzmann case, Eq. (6.21) resumes as

$$q = \int_0^\infty g(E)z\,e^{-\beta E}\,dE. \tag{C.64}$$

Considering the density of states

$$g(E) = \frac{\bar{g}V}{4\pi^2}\left(\frac{2m}{\hbar^2}\right)^{3/2}E^{1/2} \tag{C.65}$$

and the following integral

$$\int_0^\infty E^{1/2} e^{-\beta E} \, dE = \frac{\sqrt{\pi}}{2} \beta^{-3/2}, \tag{C.66}$$

it is possible to write

$$q = \frac{\bar{g}V}{4\pi^2} \left(\frac{2m}{\hbar^2}\right)^{3/2} z \frac{\sqrt{\pi}}{2} \beta^{-3/2} \tag{C.67}$$

and, consequently,

$$\begin{aligned}
\Phi &= -\frac{1}{\beta} q \\
&= -\frac{\bar{g}V}{4\pi^2} \left(\frac{2m}{\hbar^2}\right)^{3/2} z \frac{\sqrt{\pi}}{2} \beta^{-5/2}.
\end{aligned} \tag{C.68}$$

To recover the result presented in Eq. (6.41), we must recall one specific value of the Gamma function: $\Gamma(5/2) = 3\sqrt{\pi}/4$.

3. For high temperature $(T \to \infty)$ the chemical potential is (Eq. (6.43)):

$$\mu = \beta^{-1} \ln\left[\frac{N}{V}\frac{1}{\bar{g}} \left(\frac{4\pi \hbar^2 \beta}{2m}\right)^{3/2}\right] \tag{C.69}$$

and therefore, since $\beta = 1/k_B T$, it reads as

$$\mu \propto -T \ln T. \tag{C.70}$$

The fugacity is then

$$z = e^{\mu\beta} \propto e^{-\ln T} = \frac{1}{T} \tag{C.71}$$

and then $z \to 0$ for the $T \to \infty$ limit.

Chapter 7

1. The nearest peaks of the de Haas-van Alphen magnetization are at

$$\frac{M}{N}(x = n) = +\mu_B, \quad \frac{M}{N}(x = n+1) = -\mu_B. \tag{C.72}$$

Thus, the amplitude of the magnetic oscillations is $2\mu_B$.

2. The hydrogen atom was discussed in Complement 3.A and therefore the reader is able to evaluate the diamagnetic contribution of the $1s$ state. The wave function for this state is

$$\Psi_{1s} = (\pi a_0^3)^{-1/2} e^{-r/a_0}, \tag{C.73}$$

where $a_0 = 4\pi\epsilon_0\hbar^2/me^2 = 5.29 \times 10^{-11}$ m is the Bohr radius.

From Section 7.1, where we discussed the diamagnetic behavior of localized electrons, it is possible to recover the magnetization and magnetic susceptibility, that read as

$$M = -N\left(\frac{e}{3\hbar}\langle r^2\rangle\right)\mu_0 H\mu_B, \tag{C.74}$$

$$\chi_d = -N\left(\frac{e}{3\hbar}\langle r^2\rangle\right)\mu_0\mu_B. \tag{C.75}$$

Thus, we need to evaluate $\langle r^2\rangle$, and this operation can be easily done[1]

$$\langle r^2\rangle = \int r^2 |\Psi_{1s}|^2 \, d\vec{r}$$

$$= \int_0^\pi \int_0^{2\pi} \int_0^\infty r^2 (\pi a_0^3)^{-1} e^{-2r/a_0} r^2 \sin(\theta) dr \, d\phi \, d\theta \tag{C.77}$$

and therefore[2]

$$\langle r^2\rangle = 3a_0^2. \tag{C.78}$$

Following, the term inside parentheses (on the magnetization and susceptibility above), is

$$\left(\frac{e}{3\hbar}\langle r^2\rangle\right) = 4.27 \times 10^{-6} \text{ T}^{-1}. \tag{C.79}$$

Thus, the magnetization, for $\mu_0 H = 1$ T of applied magnetic field, is

$$\frac{M}{N} = -4.27 \times 10^{-6}\mu_B \tag{C.80}$$

while the magnetic susceptibility is

$$\frac{\chi_d}{N} = -4.27 \times 10^{-6}\mu_0\mu_B \text{ T}^{-1}. \tag{C.81}$$

If you are not comfortable with these units, try to change; note that $\mu_0\mu_B$ T^{-1} = 1.17×10^{-29} m^3.

---

[1] The Jacobian of this transformation is

$$d\vec{r} = dx \, dy \, dz = r^2 \sin\theta \, dr \, d\theta \, d\phi, \tag{C.76}$$

following the definitions in Figure 3.4.

[2]
$$\int_0^\infty x^4 e^{-ax} = \frac{24}{a^5}.$$

3. For high temperature and low magnetic field, consider in Eq. (7.57) the approximation $x = \mu_B B \beta \to 0$. This assumption makes possible the following approximation

$$L(x) \xrightarrow{x \to 0} \frac{x}{3} - \frac{x^3}{45} + \cdots \tag{C.82}$$

and therefore

$$M = -N \left(\frac{\mu_B}{3k_B T} \right) \mu_0 H \mu_B. \tag{C.83}$$

Consequently, the magnetic susceptibility reads as

$$\chi_d = -N \left(\frac{\mu_B}{3k_B T} \right) \mu_0 \mu_B. \tag{C.84}$$

4. Let us recall the first term of Eq. (7.28), called as $\ln \mathcal{Z}_0$:

$$\ln \mathcal{Z}_0 = 2 \frac{VeB}{h^2} \int_0^{+\infty} \int_0^{+\infty} \ln \left\{ 1 + z \exp \left[-\beta \left(\frac{p_z^2}{2m} + 2\mu_B B x \right) \right] \right\} dx \, dp_z. \tag{C.85}$$

The limit of integration on p_z was changed as $\int_{-\infty}^{\infty} \ldots dp_z \to 2 \int_0^{\infty} \ldots dp_z$ because the function to be integrated is an even function on p_z.
The first step is to change the order of the integration, i.e., $dx \, dp_z \to dp_z \, dx$

$$\ln \mathcal{Z}_0 = 2 \frac{VeB}{h^2} \int_0^{+\infty} \int_0^{+\infty} \ln \left[1 + z \exp \left(-\beta 2\mu_B B x \right) \exp \left(-\beta \frac{p_z^2}{2m} \right) \right] dp_z \, dx \tag{C.86}$$

and then consider only the inner integration, $\mathcal{I}_1$. The next step is to change the integration on p_z to the following variable: $\epsilon = p^2/2m$. Thus,

$$\mathcal{I}_1 = \int_0^{+\infty} \ln \left[1 + z \exp \left(-\beta 2\mu_B B x \right) \exp \left(-\beta \frac{p_z^2}{2m} \right) \right] dp_z$$

$$= \sqrt{\frac{m}{2}} \int_0^{+\infty} \epsilon^{-1/2} \ln \left(1 + z \exp \left(-\beta 2\mu_B B x \right) e^{-\beta \epsilon} \right) d\epsilon. \tag{C.87}$$

Now, we need to integrate by parts,[3] and the integration reads as

$$\mathcal{I}_1 = \sqrt{2m}\beta \int_0^{+\infty} \frac{\epsilon^{1/2}}{z^{-1}\exp\left(\beta 2\mu_B Bx\right)e^{\beta\epsilon}+1}\,d\epsilon. \tag{C.88}$$

Returning this result into Eq. (C.86) and changing again the order of the integrals, i.e., $d\epsilon\,dx \to dx\,d\epsilon$, we found

$$\ln \mathcal{Z}_0 = 2\sqrt{2m}\beta \frac{VeB}{h^2} \int_0^{+\infty} \epsilon^{1/2} \int_0^{+\infty} \frac{1}{z^{-1}\exp\left(\beta 2\mu_B Bx\right)e^{\beta\epsilon}+1}\,dx\,d\epsilon. \tag{C.89}$$

The inner integral $\mathcal{I}_2$ is easily solved:

$$\mathcal{I}_2 = \int_0^{+\infty} \frac{dx}{z^{-1}\exp\left(\beta 2\mu_B Bx\right)e^{\beta\epsilon}+1}$$

$$= \frac{1}{2\mu_B B\beta}\ln\left(1+ze^{-\beta\epsilon}\right). \tag{C.90}$$

Thus, Eq. (C.86) holds

$$\ln \mathcal{Z}_0 = \frac{2\pi V(2m)^{3/2}}{h^3}\int_0^{+\infty}\epsilon^{1/2}\ln[1+ze^{-\beta\epsilon}]\,d\epsilon \tag{C.91}$$

considering $\mu_B = e\hbar/2m$.

Chapter 8

1. We can write the equation

$$C_M = C_B - T\left[\frac{\partial M}{\partial T}\bigg|_B\right]^2\left[\frac{\partial M}{\partial B}\bigg|_T\right]^{-1} \tag{C.92}$$

in terms of reduced quantities defined in Chapter 8. Thus,

$$c_m = c_b - \beta^3\left[\frac{\partial m(x)}{\partial x}\frac{\partial x}{\partial\beta}\bigg|_b\right]^2\left[\frac{\partial m(x)}{\partial x}\frac{\partial x}{\partial b}\bigg|_\beta\right]^{-1}$$

$$= c_b - x^2\frac{\partial m(x)}{\partial x}. \tag{C.93}$$

---

[3] Integration by parts is based on the following relationship

$$\int u\,dv = uv - \int v\,du,$$

where, for our particular case, we considered

$$dv = \epsilon^{-1/2}\,d\epsilon$$

and

$$u = \ln\left[1+z\exp\left(-\beta 2\mu_B Bx\right)e^{-\beta\epsilon}\right].$$

From Eq. (8.24), we know that

$$c_b(x) = x^2 \frac{\partial m(x)}{\partial x} \tag{C.94}$$

and then

$$c_m = 0. \tag{C.95}$$

2. From Eqs. (8.80) and (8.59) we know, respectively

$$s(x) \xrightarrow{x \to 0} \ln(2j+1) - \frac{(j+1)}{6j} x^2 \tag{C.96}$$

and

$$m(x) \xrightarrow{x \to 0} \frac{(j+1)}{3j} x \;\Rightarrow\; x = \frac{3j}{(j+1)} m. \tag{C.97}$$

Then, for $x \to 0$

$$s(m) = \ln(2j+1) - \frac{3}{2} \frac{j}{(j+1)} m^2. \tag{C.98}$$

Now, it is easy to see that

$$C_M = T \left. \frac{\partial S}{\partial T} \right|_M = 0. \tag{C.99}$$

3. The Brillouin function is given by (Eq. (8.50)):

$$B_j(x) = a \coth(ax) - b \coth(bx), \tag{C.100}$$

where $a = 1 + b$ and $b = 1/2j$. Thus,

$$bx \xrightarrow{j \to \infty} 0 \;\Rightarrow\; \coth(bx) \xrightarrow{j \to \infty} \frac{1}{bx}. \tag{C.101}$$

In addition

$$a = 1 + b \xrightarrow{j \to \infty} 1 \tag{C.102}$$

and, consequently

$$B_j(x) \xrightarrow{j \to \infty} \coth(x) - \frac{1}{x} = L(x), \tag{C.103}$$

where $L(x)$ is the Langevin function.

4. The total magnetization can be written as the sum of the magnetization of each subgroups, "up" and "down"

$$M = M_\Uparrow + M_\Downarrow. \tag{C.104}$$

Generally speaking, the magnetization can be obtained from the *Grand* canonic potential $\Phi(T, B, z)$ and therefore

$$M_\Uparrow = -\frac{\partial}{\partial B} \Phi_\Uparrow(z') = -\frac{\partial}{\partial z'} \Phi_\Uparrow(z') \frac{\partial z'}{\partial B}, \tag{C.105}$$

where

$$z' = \exp\left[\beta(\mu_\Uparrow + \mu_B B)\right] \tag{C.106}$$

and therefore

$$\frac{\partial z'}{\partial B} = \mu_B \beta z'. \tag{C.107}$$

Thus,

$$M_\Uparrow = -\mu_B \beta z' \frac{\partial}{\partial z'} \Phi_\Uparrow(z'). \tag{C.108}$$

Analogously,

$$M_\Downarrow = \mu_B \beta z'' \frac{\partial}{\partial z''} \Phi_\Downarrow(z''), \tag{C.109}$$

where

$$z'' = \exp\left[\beta(\mu_\Downarrow - \mu_B B)\right]. \tag{C.110}$$

Let us now derive the total number of particles in each subgroup. It can also be derived from the *Grand* canonic potential:

$$N_\Uparrow = -\frac{\partial}{\partial \mu_\Uparrow} \Phi_\Uparrow(z') = -\frac{\partial}{\partial z'} \Phi_\Uparrow(z') \frac{\partial z'}{\partial \mu_\Uparrow} \tag{C.111}$$

and then

$$N_\Uparrow = -\beta z' \frac{\partial}{\partial z'} \Phi_\Uparrow(z'). \tag{C.112}$$

Analogously,

$$N_\Downarrow = -\beta z'' \frac{\partial}{\partial z''} \Phi_\Downarrow(z''). \tag{C.113}$$

Substituting $N_\Uparrow$ into $M_\Uparrow$ and $N_\Downarrow$ into $M_\Downarrow$ it is now possible to derive the magnetization of each subgroup:

$$M_\Uparrow = N_\Uparrow \mu_B \quad \text{and} \quad M_\Downarrow = -N_\Downarrow \mu_B \tag{C.114}$$

and, consequently, the total magnetization

$$M = (N_\Uparrow - N_\Downarrow)\mu_B. \tag{C.115}$$

Evaluate the magnetization as the difference in the number of magnetic moments "up" and "down" is also quite reasonable, since for zero external magnetic field, $N_\Uparrow = N_\Downarrow$ and the magnetization is zero. Application of an external magnetic field favors the migration of electrons from the "up" subgroup to the "down" one and therefore $N_\Uparrow > N_\Downarrow$. Consequently, the number of unbalanced electrons is proportional to the magnetization (in μ_B units).

5. The final expressions for $q_\Uparrow(T, B, z)$ and $q_\Downarrow(T, B, z)$ (Eqs. (8.105) and (8.109), respectively) have the same mathematical structure of that already solved for a simple electron gas (Eq. (6.21)). The general solution for the *Grand* canonic potential ($\Phi(T, B, z) = -\frac{1}{\beta}q(T, B, z)$) is (Eq. (6.26)):

$$\Phi(T, B, z) = \frac{2}{3}\frac{\bar{g}V}{4\pi^2}\left(\frac{2m}{\hbar^2}\right)^{3/2}\beta^{-5/2}\Gamma(5/2)Li_{5/2}(-z). \tag{C.116}$$

From the above, it is possible to derive the magnetization

$$M = -\frac{\partial}{\partial B}\Phi(T, B, z) \tag{C.117}$$

that, for the present case, must be the sum of the magnetization of both subbands:

$$\begin{aligned} M &= M_\Uparrow + M_\Downarrow \\ &= \frac{\partial}{\partial B}\Phi_\Uparrow(T, B, z') + \frac{\partial}{\partial B}\Phi_\Downarrow(T, B, z''). \end{aligned} \tag{C.118}$$

For the zero temperature limit, the fugacity $z = e^{\mu\beta}$ tends to infinity and therefore the PolyLog function resumes as

$$Li_n(-z) \approx -\frac{[\ln(z)]^n}{n!} \quad (z \to \infty) \tag{C.119}$$

Thus, the *Grand* canonic potential of the "up" subband is

$$\Phi_\Uparrow(T, B, z') = -\frac{2}{3}\frac{\bar{g}V}{4\pi^2}\left(\frac{2m}{\hbar^2}\right)^{3/2}\beta^{-5/2}\Gamma(5/2)\frac{[\ln(z')]^{5/2}}{(5/2)!} \tag{C.120}$$

and then[4]

$$\Phi_\Uparrow(T, B, z') = -\frac{1}{5}N\epsilon_F^{-3/2}\left(\epsilon_F + \mu_B B\right)^{5/2}. \tag{C.122}$$

---

[4]For that evaluation we have used $n! = \Gamma(n+1)$, $\Gamma(5/2) = \frac{3}{4}\sqrt{\pi}$, and $\Gamma(7/2) = \frac{15}{8}\sqrt{\pi}$. Since we are working at $T = 0$ and low magnetic field limits, we can write

$$\mu = \epsilon_F(N/2, \bar{g} = 1) = \left(\frac{N}{V}3\pi^2\right)^{2/3}\frac{\hbar^2}{2m}. \tag{C.121}$$

In addition, we are describing the *Grand* canonic potential of only one subband and therefore $\bar{g} = 1$.

From the above potential it is possible to evaluate the magnetization, that reads as

$$M_\Uparrow = -\frac{\partial}{\partial B} \Phi_\Uparrow(T, B, z') = \frac{1}{2} N \epsilon_F^{-3/2} \left(\epsilon_F + \mu_B B\right)^{3/2} \mu_B \qquad \text{(C.123)}$$

and then[5]

$$M_\Uparrow = \frac{1}{2} N \left(1 + \frac{3}{2} \frac{\mu_B B}{\epsilon_F}\right) \mu_B. \qquad \text{(C.124)}$$

Analogously, to the "down" subband:

$$\Phi_\Downarrow(T, B, z'') = -\frac{1}{5} N \epsilon_F^{-3/2} \left(\epsilon_F - \mu_B B\right)^{5/2} \qquad \text{(C.125)}$$

and

$$M_\Downarrow = -\frac{1}{2} N \left(1 - \frac{3}{2} \frac{\mu_B B}{\epsilon_F}\right) \mu_B \qquad \text{(C.126)}$$

The total magnetization resumes as

$$M = M_\Uparrow + M_\Downarrow = N \left(\frac{3}{2} \frac{\mu_B}{\epsilon_F}\right) \mu_0 H \mu_B. \qquad \text{(C.127)}$$

Finally, the paramagnetic susceptibility can be written as

$$\chi_p = N \left(\frac{3}{2} \frac{\mu_B}{\epsilon_F}\right) \mu_0 \mu_B. \qquad \text{(C.128)}$$

6. (a) Nondegenerated case:
The nonperturbed contributions are easily obtained, since they correspond to the eigenvalues of the diagonal nonperturbed Hamiltonian. Thus

$$\varepsilon_1^{(0)} = 1, \quad \varepsilon_2^{(0)} = 3, \quad \varepsilon_3^{(0)} = -2. \qquad \text{(C.129)}$$

Note we have three eigenvalues of energy and the system has 3×3 dimension. The first-order correction is also easy to obtain, since those are the expected values of $\mathcal{W}$ written on the original basis. Thus

$$\varepsilon_1^{(1)} = 0, \quad \varepsilon_2^{(1)} = 0, \quad \varepsilon_3^{(1)} = 1. \qquad \text{(C.130)}$$

The second-order correction can be evaluated from Eq. (8.160). Thus,

$$\varepsilon_1^{(2)} = \frac{1^2}{1-3} + \frac{0^2}{1-(-2)} = -\frac{1}{2}, \qquad \text{(C.131)}$$

$$\varepsilon_2^{(2)} = \frac{1^2}{3-1} + \frac{0^2}{3-(-2)} = \frac{1}{2}, \qquad \text{(C.132)}$$

$$\varepsilon_3^{(2)} = \frac{0^2}{-2-1} + \frac{0^2}{-2-3} = 0. \qquad \text{(C.133)}$$

---

[5]We are working at low temperature $\epsilon_F \gg k_B T$ and low magnetic field $\epsilon_F \gg \mu_B B$ limits and therefore $\mu_B B / \epsilon_F \to 0$. Thus, the following approximation is valid: $(1 + bx)^a = 1 + abx + \cdots$ (for the $x \to 0$ limit).

Finally, the three eigenvalues are

$$\varepsilon_1 = 1 - \frac{\lambda^2}{2}, \quad \varepsilon_2 = 3 + \frac{\lambda^2}{2}, \quad \varepsilon_3 = -2 + \lambda. \tag{C.134}$$

(b) Degenerated case:
As before, the nonperturbed contributions are

$$\varepsilon_1^{(0)} = 1, \quad \varepsilon_2^{(0)} = 1, \quad \varepsilon_3^{(0)} = -2. \tag{C.135}$$

The first-order correction is obtained diagonalizing submatrices of $\mathcal{W}$. These submatrices correspond to those of $\mathcal{H}_0$ that contain the degenerate eigenvalues. Thus

$$\varepsilon_{1,2}^{(1)} = \begin{pmatrix} 0 & 1 \\ 1 & 0 \end{pmatrix}, \quad \varepsilon_3^{(1)} = (1) \tag{C.136}$$

and then

$$\varepsilon_1^{(1)} = -1, \quad \varepsilon_2^{(1)} = 1, \quad \varepsilon_3^{(1)} = 1. \tag{C.137}$$

Note the first-order correction is enough to lift the degeneracy of the system. Finally, the second-order correction can be obtained from Eq. (8.A.10):

$$\varepsilon_1^{(2)} = \frac{0^2}{1 - (-2)} = 0, \tag{C.138}$$

$$\varepsilon_2^{(2)} = \frac{1^2}{1 - (-2)} = \frac{1}{3}, \tag{C.139}$$

$$\varepsilon_3^{(2)} = \frac{0^2}{-2 - 1} + \frac{1^2}{-2 - 1} = -\frac{1}{3}. \tag{C.140}$$

After a simple concatenation of the above results, it is possible to write the final eigenvalues of energy:

$$\varepsilon_1 = 1 - \lambda, \quad \varepsilon_2 = 1 + \lambda + \frac{\lambda^2}{3}, \quad \varepsilon_3 = -2 + \lambda - \frac{\lambda^2}{3}. \tag{C.141}$$

7. From Eq. (3.51) we know that

$$g_j = 1 + \frac{j(j+1) - l(l+1) + s(s+1)}{2j(j+1)}. \tag{C.142}$$

For g_0 there is an indetermination, i.e., $0/0$. It is possible to overcome this problem considering $j = s - l$. Thus

$$\begin{aligned} g_0 &= \frac{3}{2} + \frac{s^2 + s - l^2 - l + ls - ls}{2(s-l)(s-l+1)} \\ &= \frac{3}{2} + \frac{(s-l)(s+l+1)}{2(s-l)(s-l+1)} \\ &= \frac{3}{2} + \frac{(s+l+1)}{2(s-l+1)} \end{aligned} \tag{C.143}$$

and considering $s = l$ it reads as

$$g_0 = 2 + l = 2 + s. \tag{C.144}$$

It is now easy to see that for the trivalent Europium, the ground state has $g_0 = 5$. Now, we need to give attention to α_j, that also has an indetermination. From Eq. (8.B.19) we know that

$$\alpha_j = \frac{1}{6\zeta(2j+1)}\left[\frac{F(j+1)}{j+1} - \frac{F(j)}{j}\right] \tag{C.145}$$

and

$$F(j) = \frac{1}{j}\left[(l+s+1)^2 - j^2\right]\left[j^2 - (s-l)^2\right]. \tag{C.146}$$

It is easy to see that the second term of α_j contains the indetermination. As before, considering $j = s - l$ we can write

$$\left.\frac{F(j)}{j}\right|_{j=s-l} = \left[(l+s+1)^2 - (s-l)^2\right]\left[1 - \frac{(s-l)^2}{(s-l)^2}\right] = 0 \tag{C.147}$$

and then the indetermination is removed: this term is zero.
Now it is easy to evaluate χ_0 and it is

$$\chi_0 = \alpha_0 = \frac{1}{6\zeta}\left[(2s+1)^2 - 1\right] \tag{C.148}$$

and, for trivalent Europium, it is

$$\chi_0 = \alpha_0 = \frac{24}{3\zeta}. \tag{C.149}$$

The numerator of the above equation is the same of that, the first term of Eq. (8.B.22).

Chapter 9

1. Consider the Exercise 1 of Chapter 12. We know that there are four spin wave functions for a dimer of spin $s_1 = s_2 = 1/2$. These are

$$\begin{aligned}
|1, 1\rangle &= |+, +\rangle, \\
|1, 0\rangle &= \frac{1}{\sqrt{2}}(|+, -\rangle + |-, +\rangle), \\
|1, -1\rangle &= |-, -\rangle, \\
|0, 0\rangle &= \frac{1}{\sqrt{2}}(|+, -\rangle - |-, +\rangle).
\end{aligned} \tag{C.150}$$

The states on the left are written in the coupled basis $|s, m_s\rangle$, while those on the right are the same states, however, written in the local basis $|m_{s1}, m_{s2}\rangle$ ("+" and "−" means spin up and spin down, respectively).

Only one of those states has an antisymmetric wave function ($|0, 0\rangle$). This state has $s = 0$ and then antiparallel alignment of the spins. The other three have a symmetric character and $s = 1$; consequently, a parallel alignment of the spins. Remember, we are dealing with electrons and therefore, fermions. Thus, the total wave function (spatial times spin) must be antisymmetric. To satisfy this condition, spatial symmetric (antisymmetric) wave function must multiply antisymmetric (symmetric) spin wave function.

We finally conclude the aim of this exercise: spatial symmetric wave function leads to antisymmetric spin wave function that, on its turn, has an antiparallel alignment of the spins. Analogously, spatial antisymmetric wave function leads to symmetric spin wave function (that has a parallel alignment).

2. First, the local magnetocrystalline anisotropy is given by Eq. (9.69):

$$\mathcal{H}_{ani} = \vec{S} \cdot \overleftrightarrow{\mathcal{D}}_{ani} \cdot \vec{S} \tag{C.151}$$

that can be rewritten in accordance with Eq. (9.96):

$$\mathcal{H} = D\left(S_z^2 - \frac{1}{3}S^2\right) + E\left(S_x^2 - S_y^2\right). \tag{C.152}$$

Following, we need the spin operators (see Section 3.1)

$$S_x = \frac{1}{2}\begin{pmatrix} 0 & 1 \\ 1 & 0 \end{pmatrix}, \quad S_y = \frac{i}{2}\begin{pmatrix} 0 & -1 \\ 1 & 0 \end{pmatrix}, \quad S_z = \frac{1}{2}\begin{pmatrix} 1 & 0 \\ 0 & -1 \end{pmatrix}. \tag{C.153}$$

Then, after some simple algebraic operations, we obtain

$$\mathcal{H} = \begin{pmatrix} 0 & 0 \\ 0 & 0 \end{pmatrix}. \tag{C.154}$$

It means that $s = 1/2$ spin does not have local magnetocrystalline anisotropy.

Chapter 11

1. We know from Eq. (11.40) that

$$\xi_{+-} = \left(\frac{g_4 - \sqrt{g_4^2 - 4g_2g_6}}{2g_6}\right)^{1/2}. \tag{C.155}$$

Thus, $\xi_{+-} = 0$ leads to $g_2 = 0$ and then $T = T_0$.

Chapter 12

1. This exercise considers the simplest dimer: $s_1 = s_2 = 1/2$. The coupled basis requires

$$|s_1 - s_2| \le s \le s_1 + s_2, \tag{C.156}$$

and then $0 \le s \le 1$. Thus, the eigenstates are

$$\text{local basis } (|m_{s1}, m_{s2}\rangle): \left\{ \begin{array}{c} |+, +\rangle \\ |+, -\rangle \\ |-, +\rangle \\ |-, -\rangle \end{array} \right\}, \quad \text{coupled basis}(|s, m_s\rangle): \left\{ \begin{array}{c} |1, 1\rangle \\ |1, 0\rangle \\ |1, -1\rangle \\ |0, 0\rangle \end{array} \right\}.$$

Note "+" and "−" are, respectively, $+1/2$ and $-1/2$.

The Clebsch-Gordan coefficients measure the degree of correlation between eigenstates of each basis. Since there are four eigenstates for each basis, there must be 16 coefficients, providing a relation between all local eigenstate with all coupled basis. In this sense, the final result is a matrix where the lines are *bras* ($\langle \cdots |$) of the coupled basis and rows are *kets* ($| \cdots \rangle$) of the local basis; and each element is a Clebsch-Gordan coefficient measuring a specific correlation. Thus:

$$\begin{pmatrix} \langle 1, 1|+, +\rangle & \langle 1, 1|+, -\rangle & \langle 1, 1|-, +\rangle & \langle 1, 1|-, -\rangle \\ \langle 1, 0|+, +\rangle & \langle 1, 0|+, -\rangle & \langle 1, 0|-, +\rangle & \langle 1, 0|-, -\rangle \\ \langle 1, -1|+, +\rangle & \langle 1, -1|+, -\rangle & \langle 1, -1|-, +\rangle & \langle 1, -1|-, -\rangle \\ \langle 0, 0|+, +\rangle & \langle 0, 0|+, -\rangle & \langle 0, 0|-, +\rangle & \langle 0, 0|-, -\rangle \end{pmatrix}. \tag{C.157}$$

The null elements are easy to be identified, since they do not satisfy the condition of Eq. (3.32). The nonzero elements can be obtained from Eq. (3.21), that, on its turn, can be obtained from Racah formula (Eq. (3.32)). Note it is hard to be obtained by hand, but the *Cardamomo* package (see Section 12.1.3) provides these coefficients automatically. Finally, the matrix to change the basis is

$$U = \begin{array}{cccc} |+, +\rangle & |+, -\rangle & |-, +\rangle & |-, -\rangle \\ \begin{pmatrix} 1 & 0 & 0 & 0 \\ 0 & 1/\sqrt{2} & 1/\sqrt{2} & 0 \\ 0 & 0 & 0 & 1 \\ 0 & 1/\sqrt{2} & -1/\sqrt{2} & 0 \end{pmatrix} & \begin{array}{c} \langle 1, 1| \\ \langle 1, 0| \\ \langle 1, -1| \\ \langle 0, 0| \end{array} \end{array} \tag{C.158}$$

and the basis change occurs as in Eq. (3.20), that can be rewritten as

$$|s, m_s\rangle = U |m_{s1}, m_{s2}\rangle. \tag{C.159}$$

The relations to change the basis resume as

$$|1, 1\rangle = |+, +\rangle,$$

$$|1, 0\rangle = \frac{1}{\sqrt{2}} \left(|+, -\rangle + |-, +\rangle \right),$$

$$|1, -1\rangle = |-, -\rangle,$$

$$|0, 0\rangle = \frac{1}{\sqrt{2}} \left(|+, -\rangle - |-, +\rangle \right), \tag{C.160}$$

or

$$|+, +\rangle = |1, 1\rangle,$$

$$|+, -\rangle = \frac{1}{\sqrt{2}} (|1, 0\rangle + |0, 0\rangle),$$

$$|-, +\rangle = \frac{1}{\sqrt{2}} (|1, 0\rangle - |0, 0\rangle),$$

$$|-, -\rangle = |1, -1\rangle. \tag{C.161}$$

2. A dimer with dipolar (Eq. (9.35)) and isotropic Heisenberg (Eq. (9.21)) interaction has the following Hamiltonian

$$\mathcal{H} = -J\vec{S}_1 \cdot \vec{S}_2 + \vec{S}_1 \cdot \overleftrightarrow{\mathcal{D}}_d \cdot \vec{S}_2, \tag{C.162}$$

where, in accordance with what was defined to the tensor $\overleftrightarrow{\mathcal{D}}$ (see Section 9.5), the above Hamiltonian can be rewritten as

$$\mathcal{H} = -J\vec{S}_1 \cdot \vec{S}_2 + D_d \left(S_{1z} S_{2z} - \frac{1}{3} \vec{S}_1 \cdot \vec{S}_2 \right) + E_d \left(S_{1x} S_{2x} - S_{1y} S_{2y} \right). \tag{C.163}$$

The matrix form of this Hamiltonian is

$$\mathcal{H} = \begin{pmatrix} D_d/6 - J/4 & 0 & E_d/2 & 0 \\ 0 & -D_d/3 - J/4 & 0 & 0 \\ E_d/2 & 0 & D_d/6 - J/4 & 0 \\ 0 & 0 & 0 & 3J/4 \end{pmatrix} \begin{matrix} \langle 1, 1| \\ \langle 1, 0| \\ \langle 1, -1| \\ \langle 0, 0| \end{matrix}, \tag{C.164}$$

where the spin operators were built up in the coupled basis $|s, m_s\rangle$ and the matrix organized following the states given on the side of the matrix. To evaluate the magnetic susceptibility, first we need to excite the system with an external magnetic field, given by the Zeeman term:

$$\mu_B \vec{B} \cdot \overleftrightarrow{g} \cdot \vec{S} = \mu_B \sum_u \sum_v g_{uv} S_u B_v, \tag{C.165}$$

where, considering only the diagonal terms of the $\overleftrightarrow{g}$ tensor, the Zeeman term resumes as

$$\mu_B \sum_u g_{uu} S_u B_u, \tag{C.166}$$

where $u = x, y, z$. It is possible to consider the magnetic field along an arbitrary direction and for this case, the Zeeman term is

$$\mathcal{H} = \begin{pmatrix} b_z & b_-/\sqrt{2} & 0 & 0 \\ b_+/\sqrt{2} & 0 & b_-/\sqrt{2} & 0 \\ 0 & b_+/\sqrt{2} & -b_z & 0 \\ 0 & 0 & 0 & 0 \end{pmatrix}, \tag{C.167}$$

where

$$b_+ = b_x + i b_y, \tag{C.168}$$
$$b_- = b_x - i b_y, \tag{C.169}$$

and

$$b_u = g_u B_u \mu_B. \tag{C.170}$$

Now, it is possible to do some simplifications. First, consider only the axial contribution, i.e., $E_d = 0$. For this case, the Hamiltonian of Eq. (C.164) is diagonal and then it is possible to use perturbation theory to obtain the eigenvalues of energy. Following the possible approximations to be done, let us consider the magnetic field along z (and therefore $b_x = b_y = 0$), and the energy eigenvalues can be obtained (see Table C.1). Next step is to evaluate the eigenvalues for the case in which the magnetic field is along x (see again Table C.1). Details of how to obtain the eigenvalues are in Complement 8.A, that describes the Perturbation Theory; however, the *Cardamomo* package handles it (see Section 12.13). With the energy eigenvalues, it is possible to evaluate the van Vleck susceptibility. Thus

$$\chi_x = N_d \mu_0 \frac{4(g_x \mu_B)^2}{D_d} \frac{e^{x/3} - e^{-x/6}}{2e^{-x/6} + e^{x/3} + e^{-y}} \tag{C.171}$$

Table C.1 Energy eigenvalues considering the dipolar axial interaction ($D_d \neq 0$ and $E = 0$) and two cases of applied magnetic field: along (i) z and (ii) x. The case in which the magnetic field is along y returns the same eigenvalues of x. This table is for a dimer with $s_1 = s_2 = 1/2$.

$\lvert s, m_s \rangle$	$E_{m_s}^{(0)}$	$E_{m_s}^{(1)}(b_z)$	$E_{m_s}^{(2)}(b_z)$	$E_{m_s}^{(1)}(b_x)$	$E_{m_s}^{(2)}(b_x)$
$\lvert 1, 1 \rangle$	$-J + D_d/6$	b_z	0	0	b_x^2/D_d
$\lvert 1, 0 \rangle$	$-J - D_d/3$	0	0	0	$-2b_x^2/D_d$
$\lvert 1, -1 \rangle$	$-J + D_d/6$	$-b_z$	0	0	b_x^2/D_d
$\lvert 0, 0 \rangle$	0	0	0	0	0

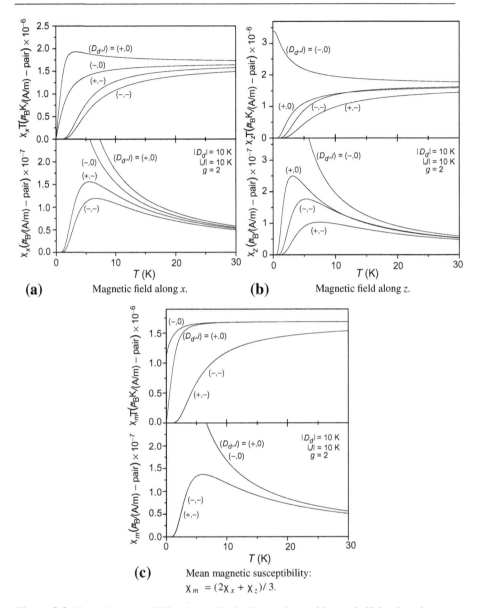

Figure C.3 Magnetic susceptibility due to dipolar interaction and isotropic Heisenberg interaction. Susceptibility times temperature (upper panels) and susceptibility (bottom panels), when the magnetic field is along x and z; also shown is the average magnetic susceptibility.

and

$$\chi_z = N_{\mathrm{d}}\mu_0 \frac{2(g_z\mu_B)^2}{k_B T} \frac{e^{-x/6}}{2e^{-x/6} + e^{x/3} + e^{-y}}, \qquad (\mathrm{C}.172)$$

where

$$x = \frac{D_d}{k_\mathrm{B} T},\qquad\qquad\qquad\qquad\text{(C.173)}$$

$$y = \frac{J}{k_\mathrm{B} T},\qquad\qquad\qquad\qquad\text{(C.174)}$$

and N_d is the number of dimers in the system. Finally, the average susceptibility (for a polycrystal) is

$$\chi_m = \frac{1}{3}(\chi_z + 2\chi_x).\qquad\qquad\qquad\qquad\text{(C.175)}$$

The behavior of this susceptibility for the cases $D_d > 0$ and $D_d < 0$ is given in Figure C.3, considering the case in which $J < 0$ and $J = 0$ (this last is obviously a pure dipolar interaction). Note, considering a polycrystal, i.e., micro-crystals along all of the directions, the average susceptibility χ_m (Eq. (C.175)) is the same for both cases: $D_d > 0$ and $D_d < 0$. Thus, for a polycrystal, it is difficult to determine the sign of the parameter D_d. On the other hand, there is a huge difference dealing with a single crystal; for $D_d > 0$ (and $J = 0$), $\chi_z T$ goes to zero decreasing temperature and, on the contrary, for $D_d < 0$ (and $J = 0$), $\chi_z T$ increases by decreasing temperature.

This page is intentionally left blank

Bibliography

[1] Jensen J, Mackintosh A. Rare earth magnetism: structures and excitations. Oxford: Clarendon Press; 1991.

[2] Nolting W, Ramakanth A. Quantum theory of magnetism. Heidelberg: Springer; 2009.

[3] van Vleck J. The theory of electric and magnetic susceptibilities. Oxford: Clarendon Press; 1932.

[4] Takikawa Y, Ebisu S, Nagata S. J Phys Chem Solids 2010;71:1592.

[5] Moryia T. Phys Rev 1960;120:91.

[6] Dzialoshinsky I. J Phys Chem Solids 1958;4:241.

[7] Reis MS, Amaral VS, Araújo JP, Tavares PB, Gomes AM, Oliveira IS. Phys Rev B 2005;71:144413.

[8] Warburg E. Ann Phys (Leipzig) 1881;13:141.

[9] Debye P. Ann Phys 1926;81:1154.

[10] Giauque W. J Am Chem Soc 1927;49:1864.

[11] Giauque WF, MacDougall DP. Phys Rev 1933;43:768.

[12] <http://www.universe.nasa.gov/xrays/programs/astroe/eng/adr.html>.

[13] Gschneidner K, Pecharsky VK, Tsokol A. Rep Prog Phys 2005;68:1479.

[14] Kuhn L, Pryds N, Bahl C, Smith A. J Phys: Conf Ser 2011;303:012082.

[15] Reis MS, Rubinger RM, Sobolev NA, Valente MA, Yamada K, Sato K, et al. Phys Rev B 2008;77:104439.

[16] Pecharsky VK, Gschneidner Jr KA. Phys Rev Lett 1997;78:4494.

[17] Choe W, Pecharsky VK, Pecharsky AO, Gschneidner KA, Young VG, Miller GJ. Phys Rev Lett 2000;84:4617.

[18] Pecharsky VK, Holm AP, Gschneidner KA, Rink R. Phys Rev Lett 2003;91:197204.

[19] Morellon L, Arnold Z, Magen C, Ritter C, Prokhnenko O, Skorokhod Y, et al. Phys Rev Lett 2004;93:137201.

[20] Yamada H, Goto T. Phys Rev B 2003;68.

[21] Fujita A, Fujieda S, Hasegawa Y, Fukamichi K. Phys Rev B 2003;67:104416.

[22] Fujita A, Fujieda S, Fukamichi K, Mitamura H, Goto T. Phys Rev B 2001;65:014410.

[23] Wada H, Tanabe Y. Appl Phys Lett 2001;79:3302.

[24] Gama S, Coelho AA, de Campos A, Carvalho AMG, Gandra FCG, von Ranke PJ, et al. Phys Rev Lett 2004;93:237202.

[25] Hu F, Shen B, Sun J. Appl Phys Lett 2000;76:460.

[26] Zhou X, Li W, Kunkel HP, Williams G. J Phys: Condens Matter 2004;16:L39.

[27] Morelli D, Mance A, Mantese J, Micheli A. J Appl Phys 1996;79:373.

[28] Amaral V, Amaral J. J Magn Magn Mater 2004;272:2104.

[29] Chen H, Lin C, Dai D. J Magn Magn Mater 2003;257:254.

[30] Wang D, Liu H, Tang S, Yang S, Huang S, Du Y. Phys Lett A 2002;297:247.

[31] de Oliveira NA, von Ranke PJ, Troper A. Phys. Rev. B 2004;69:064421.

[32] von Ranke PJ, Nóbrega EP, de Oliveira IG, Gomes AM, Sarthour RS. Phys Rev B 2001;63:184406.

[33] Dan'kov SY, Ivtchenko VV, Tishin AM, Gschneidner KA, Pecharsky VK. Adv Cryog Eng 2000;46:397.

[34] Canepa F, Napoletano M, Cirafici S. Intermetallics 2002;10:731.

[35] Niu X, Gschneidner K, Pecharsky A, Pecharsky V. J Magn Magn Mater 2001;234:193.

[36] Brandao P, Rocha J, Reis MS, dos Santos AM, Jin R. J Solid State Chem 2009;182:253.

[37] Bogani L, Wernsdorfer W. Nat Mater 2008;7:179.

[38] Souza AM, Soares-Pinto DO, Sarthour RS, Oliveira IS, Reis MS, Brando P, et al. Phys Rev B 2009;79:054408.

[39] Manoli M, Johnstone RDL, Parsons S, Murrie M, Affronte M, Evangelisti M, Brechin EK. Angew Chem Int Ed 2007;46:4456.

[40] Hashimoto K, Ohkoshi H. Philos Trans R Soc Lond A 1999;357:2977.

[41] <http://profs.if.uff.br/marior/mm>.

[42] Reis MS. Comput Phys Commun 2011;182:1169.

[43] Reis MS. Comput Phys Commun 2012;183:99.

[44] Reis MS, dos Santos AM, Amaral VS, Brandão P. Rocha. J Phys Rev B 2006;73:214415.

[45] Bonner JC, Fisher ME. Phys Rev 1964;135:A640.

[46] Meyer A, Gleizes A, Girerd J, Verdaguer M, Kahn O. Inorg Chem 1982;21:1729.

[47] Fisher M. Am J Phys 1964;32:343.

[48] Kramers H, Wannier G. Phys Rev 1941;60:252.

[49] Fisher M. J Math Phys 1963;4:124.

[50] Drillon M, Coronado E, Georges R, Giangerzzo J, Curely J. Phys Rev B 1989;40:10992.

[51] Georeges R, Borras-Almenar J, Coronado E, Curely J, Drillon M. One-dimensional magnetism: an overview of the models. In: Magnetism: molecules to materials I: models and experiments. Wiley-VCH Verlag GmbH; 2002. p. 1.

[52] Drillon M, Coronado E, Beltran D, Georeges R. J Appl Phys 1985;57:3353.

[53] Drillon M, Coronado E, Beltran D, Georges R. Chem Phys 1983;79:449.

[54] Duffy W, Barr K. Phys Rev 1968;165:647.

[55] Hall J, Marsh W, Weller R, Hatfield W. Inorg Chem 1981;20:1033.

[56] Borras-Almenar JJ, Coronado E, Curely J, Georges R, Gianduzzo JC. Inorg Chem 1994;33:5171.

[57] Shi FN, Reis MS, Brando P, Souza AM, Flix V, Rocha J. Transit Met Chem 2010;35:779.

[58] Friedman J, Sarachik M, Tejada J, Ziolo R. Phys Rev Lett 1996;76:3830.

[59] Tupitsyn I, Barbara B. Quantum tunneling of magnetization in molecular complexes. In: Magnetism: molecules to materials III: models and experiments. Wiley-VCH Verlag GmbH; 2002. p. 109.

[60] Huang K. Statistical Mechanics. John Wiley and Sons, Inc., 1928.

Index

Printed and bound by CPI Group (UK) Ltd, Croydon, CR0 4YY

03/10/2024

01040425-0006